Springer Japan KK

Fumikazu Yoshida

The Economics of Waste and Pollution Management in Japan

With 23 Figures

 Springer

Fumikazu Yoshida
Professor
Graduate School of Economics
Hokkaido University
Sapporo 060-0809, Japan

This book was financially supported by the Japan Society for the Promotion of Science (Grant-in-Aid for Publication of Scientific Research Results: Grant No. 135252) and the Suntory Foundation (support for overseas publication).

ISBN 978-4-431-67034-6

Library of Congress Cataloging-in-Publication Data

Yoshida, Fumikazu, 1950–
    The economics of waste and pollution management in Japan / Fumikazu Yoshida.
        p.   cm.
    Includes bibliographical references (p.) and index.
    ISBN 978-4-431-67034-6   ISBN 978-4-431-67032-2 (eBook)
    DOI 10.1007/978-4-431-67032-2

    1. Refuse and refuse disposal—Japan—Costs.   2. Water quality
management—Japan—Costs.   3. Air quality management—Japan—Costs.   4. Soil
pollution—Japan—Costs.   5. Pollution—Japan—Costs.   6. Environmental policy—Economic
aspects—Japan.   I. Title.

HD4485.J3 Y85 2001
338.4'33637'00952—dc21

                                                                                          2001054257

Printed on acid-free paper

Typesetting: Best-set Typesetter Ltd., Hong Kong

SPIN: 10847349

Preface

On the threshold of the twenty-first century, the ordinary Japanese citizen has become acutely aware that various environmental hazards pose a serious threat to daily life; such hazards include the problems of waste disposal, dioxin and other substances that disturb humans' endocrine balance.

Who, a mere decade ago when these problems were first brought to our attention, would have anticipated that these environmental problems would so quickly become so common and so serious? At the same time, environmental problems on a global scale, such as ozone depreciation by chlorofluorocarbons (CFCs), global warming and climate change, have become topics of everyday conversation.

The main objective of this book is to take in these environmental problems, focusing on the two locally important and interrelated issues of waste and pollution. This will enable us to investigate the whole range of problems, from regionally based pollution caused by waste disposal and dioxin to the transboundary warming brought about by CO_2 and CFCs. We shall thus be able to analyze comprehensively the whole extent of "waste and pollution" problems, ranging from those caused by real garbage and domestic waste to the many kinds of technologically generated waste that result from the production, circulation, and consumption of industrial goods and services.

A second objective of this book is to try, not only to inquire into and suggest technical solutions to these problems, but also to evaluate the various kinds of systematic social reform that may also be necessary. To undertake this analysis, I shall use the methodology of political economy. This approach focuses, in the first place, on economic activity within the broad spectrum of the legal and political system. It also makes considerable use of historical analysis in order to achieve a wider perspective while, at the same time, engaging in comparative analysis. By these means, we can set our present-day problems in historical context and, perhaps, thereby find a clue that will lead to their solution.

Furthermore, because this book takes special note of the "Relationship between Technology and the Environment", I have attempted to scrutinize the experiences in this regard of Europe and the United States of America (USA), and have tried to show how this, in turn, relates to what has happened in the countries of Asia. I wish to underline the importance of a long range and a wide view.

The book consists of eight related chapters that deal with these issues as they affect both the global community and the particular culture and society of Japan.

The introduction defines the nature of waste and explores the problems of recycling by concentrating on the fundamental problems that the economic system must deal with in the face of mass production, mass consumption, and mass waste.

Chapter 1 analyzes methods by which environmental cleanup and cost-bearing factors are compared and assessed as the theoretical foundation for the economic analysis.

Chapter 2 examines environmental politics in Japan after World War II as a means to understand the background of present-day Japanese environmental policy.

Chapter 3 offers an economic analysis of the conduct of waste management in Japan by focusing on waste-disposal sites, the packaging of waste, dioxin pollution, and the means of financing waste management.

Chapters 4 and 5 undertake a comparative study of the problems of high-technology pollution in both the USA and Japan as an example of the relationship between high technology and the state of the environment.

Chapter 6 further explores the technological and societal cleanup systems that have come into being in response to the geo-pollution caused by high-technology and other industries.

Chapters 7 and 8 describe the causes and consequences of Itai-Itai disease as an example of accumulated pollution caused by mining operations and the attendant obligations for cleanup that followed on from the legal findings in the Itai-Itai case.

I hope that the reader will thereby gain an empirical, as well as a theoretical, understanding of the economic aspects of waste and pollution problems found in present-day Japan.

Contents

Introduction: What Is Waste?

1. Waste and Pollution/Environmental Problems

When we consider problems of the global environment, the central category that must be analyzed first is the issue of waste. Because many sorts of environmental problems occur both at the input of metabolism, or at the interface between humans and nature, and also at the output of waste from human society, the problem of waste is one of the main issues we need to address.

From industrial waste to household waste, we find many problems associated with waste management: difficulties in siting disposal facilities, the safety of the disposal facility, dioxin emissions from incinerators of household and industrial waste, increasing disposal costs, the transboundary movement of hazardous waste, and the dangers of radioactive waste. It is almost as if society is being buried alive by polluted waste generated by the waste disposal process. In terms of global warming, CO_2 is a sort of waste in the wider meaning of the term: the emission of pollutants as a result of energy usage related to production and consumption. We can therefore perceive that the majority of global environmental problems can be analyzed as problems of waste and the pollution for which waste is responsible.

2. The Many Kinds of Waste

Types of waste can be classified according to the cause generating the waste as industrial waste generated by human production (designated as such by law because of its hazards and massiveness), and waste generated mainly by consumption (i.e., household waste).

Some types of waste are classified according to the relatively long life of the original products, for instance, construction waste, obsolete vehicles, end-of-life vehicles (ELVs), and waste electrical appliances.

Because the production of material goods is linked to their consumption, the distinction between production and consumption is relatively well defined.

Looked at like this, we see that the way in which waste is generated and what kind of waste it is depends on the production activity of society itself and on its consumption habits, not to mention society's industrial and consumption structures, its distribution system, its legal system, and its cultural customs. The production and industrial structure is particularly important, and our analysis must be careful to relate this structure to the distribution and consumption structures. For example, the problem of plastic waste is significant because of its relationship with the petrochemical industry. Thus, it becomes essential to the design of production processes that we be conscious of any consequential effects of waste and pollution.

At the same time, specific problems of waste are linked together in the chain of production→distribution→consumption→collection→treatment→disposal. (In juristic terms, the collection and transportation of waste is considered "collection", whereas intermediate treatment and final disposal are termed "disposal"; the whole process is called "treatment".)

3. "Waste" as a Relative Category and "Zero Emission"

When considered in contrast with products, waste is a relative category. If the human production process is accepted as an activity for the acquisition of objects, then waste is what is generated and has to be disposed of without further usage, whereas products are objects to be acquired. It does not follow, however, that all disposed effects are waste, because it is necessary to define waste not only according to the subjective intention of the possessors, but also according to objective criteria. According to a commentary on the Japanese Waste Disposal and Cleansing Law, waste is defined as ". . . obsolete scrap if the possessor cannot himself use it nor trade with it for money".[1] This definition is not sufficient to prevent those who wish to do so from escaping the application of the Waste Disposal and Public Cleansing Law by evading the fact that waste is often a valuable commodity.

Because the category of waste is relative, there are many examples of the reuse of waste and the materialization of waste in the composition of other products.

The history of the chemical and metal industries offers many typical examples. In this context, it would not be wrong to define "waste", *Abfäll*, as "residues", *Rückstände*. It is therefore necessary to analyze why the "bads" that are disposed of become "goods". When reused or recycled, "bads" become "goods" in terms both of economy and technology.

It is of note in the context of the relativity of waste that the United Nations has initiated a project under the heading Zero Emission. The project is intended to set up a new industry group to manage the perfect use of waste: this would involve a 100% recovery of ink from paper, the use of short fiber, and a combination of beer brewing and fish culture.

The relativity of waste implies the necessity of placing waste in the context of industry as a whole, for the overall acquisition of waste. A typical example of the whole-scale use of by-products would be if members of the chemical industry were to work together in cooperation. The Zero Emission project uses a new approach to solving the problem of the hazards of the organic material industry.

4. Fundamental Problems of Waste and Pollution

It is necessary to note that the generation of waste and the emission of waste into the environment are not one and the same, and that the decisive objective must be to ensure not only that the technology for treating emissions will lead to the reduction of waste, but that the materials generated will not harm the environment. This will be decided by socio-economic conditions. A pollution control technology ("end-of-pipe technology") is one such technological method, and it depends on factors such as the profit sought by the company, the price of materials and fuel, and the legal regulations imposed by local authorities.

It is possible to use waste as a by-product and reuse or recycle it to make other products for profit, thus saving materials and fuel by reducing the amount of pure waste. The mining industry, for example, has responded to demand by beginning to use abandoned mineral resources. After World War II, the petrochemical industry recovered waste as a by-product and began to reuse it commercially. The paper and pulp industry recovered black liquor and reused it as fuel. Thus, it is possible to take waste and turn it into a valuable commodity. All of this, however, will depend on the cooperation of intelligent and sympathetic management, the requirements laid down by law, and the concerted action of concerned citizens.

The fundamental aim is to reform production processes so that the emission of chemicals such as CFCs will be brought to an end, and, by changing over to less harmful materials and auxiliary materials, to ensure cleaner methods of production.

In cases in which materials are to be treated to prevent harmful effects, it is important to take into account the features of all the forms of waste that accumulate at the site of waste generation. It is necessary to make an assessment of the life cycle of the goods, and to analyze the amount of waste generated, as well as to consider the possibility of recycling or the likelihood of decomposition.

5. Problems of Recycling

When, in our efforts to solve the waste problem, we consider the 3Rs—reduction, reuse, and recycling, in that order—we must examine the issue of recycling with particular care.

The original aim of recycling was to reduce waste, save materials, and conserve energy. However, the promotion of recycling alone, without consideration of other factors, can lead to pollution, as in the instance of the treatment of lead batteries and by the recycling process of already treated materials.

It is necessary, therefore, to avoid having any single-minded aim for recycling: we must make recycling a tool for waste reduction, and must not only change the materials used and the production process, but also pay proper attention to an assessment of the life cycles of the products.

As an example, the number of vending machines in Japan increased fivefold between 1983 and 1993, and so, too, was there a marked increase in the number of aluminum cans. Although the rate of recycling increased from 40% in 1983 to 58% in 1993, the net consumption of aluminum cans (total consumption less recycling) between 1983 and 1993 increased threefold. By increasing the rate of collection of used cans, the manufacturers of aluminum cans have been able to expand their operations and make more cans.

In 1996, approximately 70% of collected cans were recycled as cans for reuse, while others were used for the deoxidation of die cast and in the manufacture of iron and steel. In other words, not all recycled cans are recycled as aluminum cans; steel cans may be processed at an electrical furnace into construction material.

At the same time, we have to bear in mind the extraordinary amount of energy required to operate vending machines. Japan, at present, is home to three million vending machines, and these machines consume the amount of electricity produced by a power station capable of generating one million kWh.

We must also recognize that the use of aluminum is not itself without serious side issues. The extraction of bauxite and fly ash causes environmental problems, and a relationship has been suggested between the use of aluminum tableware and Alzheimer's disease.

Should a recycling system be set up, it may become a mass operation and would, therefore, not act as a deterrent to mass production and mass consumption—and, so, might only serve to exacerbate the problem. In addition, although the saving of resources as a result of recycling may seem more likely to lead to a reduction in the use of new resources than would be the case if goods were not recycled, it does not, in fact, follow that the rate of reduction of new resources thanks to the recycling process will lead to a reduction in the input of virgin material overall.[2,3]

The modern habit of mass production and mass consumption makes effective mass disposal of waste a precondition of a healthy society. Among the instruments (or "tools") of mass production—the artifacts created by mass production for the dissemination of mass-produced goods—are packaging and other disposable (or throwaway) parts; the general public is now becoming sensitive to the problems that these disposable parts generate, and seeks for them to be regulated. Yet, regulations regarding generated hazardous waste and pollution are bound to have limits and, if the regulations regarding input and production are not addressed in the first place, the users of goods will tend to find loopholes when it comes to the output of waste. For this reason, we must first address the issue of input: the use of raw materials and their production. This will entail changes to industrial structures and production processes, changes in the raw material to be manufactured, improvements in operation and maintenance, the assessment of product life cycles, and regulation of the amount of goods mass produced.

6. Analysis of Value in Itself and Value in Use

Political economy analyzes commodities in terms of their trading value and their value in use. Such a distinction is useful when analyzing the reuse and recycling of a commodity. A commodity's "value in use" is its practical usefulness, whereas its exchange value is the rate at which a commodity is exchanged for another commodity. The essence of the exchange value is the commodity's trading value.

When defining waste, it is not enough to define it as "trading with payment" in terms of the trading value of the commodity: it is necessary also to distinguish the term "obliged to treat and dispose" from the commodity's "value in use". Such discrimination, if made into a definition, would serve to prevent traders' evasion of the waste issue by claiming (as they do) that something is not waste because it is traded with payment.

A commodity without **value** in use has no value in itself, whereas even a value in use has no trading value or may even have a negative value. If this seems strange, a typical example might be reimbursement in the recycling of used paper. In this case, even if the paper has value in use and is traded with reimbursement, the charging system on business garbage and the oversupply of used paper means that used paper has a negative price; that is, the value of the paper material minus the cost of transportation and treatment. In this case, too, the conditions of recycling—massiveness, market, stability of supply, and the technology necessary to recycle the commodity—will, together, decide the value in use and the absolute value of the recycled commodity.

Waste has value as a commodity when the disposed commodity is discovered to have value in use. There are many examples. Since World War II, the petrochemical industry has combined to recover waste as a by-product and has developed new commodities from it. Recently, the cement industry has substituted coal ash for the clay and flue gas desulfurization plaster previously used as material for cement, and has made use of the used tires as auxiliary fuel.

Although it is necessary for recycled material to be cheaper than virgin material as the price condition of the market at a certain time if the material is to be thought of as having value in itself, nonetheless, the price level of the recycled material will be dependent on the price level of the virgin material. For example, because of the appreciation in the value of the yen, the price of imported virgin material is decreasing, and the price of recycled material is also falling.[4,5] For the same reason, flue gas desulfurization plaster now has to compete with gypsum plaster.

7. Japan's Material Balance

The White Paper on the Environment (1997)[6] defined and calculated the balance between the input of resources into Japan's economic activity and the emission of obsolete scrap. *The White Paper on the Environment* calculated that Japan had extracted 2 billion tons of resources from Japan itself and from abroad. At the same time, Japan emitted 0.8 billion tons through disposal and export. Approximately 0.2 billion tons was recycled into the input. Total household and industrial waste amounted to approximately 0.45 billion tons, while CO_2 emission came to 1.2 billion tons. It is worth noting that the construction industry, which was responsible for 0.9 billion tons, took 70% of its domestic accumulated material, which gives this industry a significant role in the matter of waste and its problems. Japan, however, has to import large quantities of both unrenewable and renewable resources: in 1994, for example, Japan's iron imports amounted to 19% of world production, whereas imports of crude oil came to 8% and imports of forestry products amounted to a massive 26% of world production.

This reveals very clearly that the environmental burden that Japan imposes on the countries that supply it with raw materials deeply affects the sustainable development of the world as a whole. In addition, it shows how necessary it is that Japan's industrial and consumption structures should be placed at the center of issues and problems that affect the global environment.

Guenter Pauri, a leading proponent of the "zero-emission" of waste scheme, has pointed out that, because Japan is the biggest importer of semiprocessed material as well as being the greatest producer of industrial waste, Japan is responsible for perfecting the proper disposal of waste.[7]

The White Paper on the Environment calculated that approximately 2.5 billion tons of Japanese waste was disposed of as "a hidden flow" abroad.[8] German economists have called this hidden flow *material-Intensität Pro Service einheit* (MIPS), i.e., material intensity per service unit, and have argued that the way to reduce this MIPS is "Factor 10", which means the increasing of both labor and resource productivity.[9-11] In 1997, Japan's Central Council of the Environment produced the paper *How to Reduce the Environmental Burden Created by Waste Disposal*,[12] which analyzed the qualitative aspects of waste disposal and environmental pollution. The paper pointed out that mercury, cadmium, and lead are used in batteries, and that CFCs are used by vehicles. I would also stress that mercury, cadmium, and nickel are used in unreturnable open systems (such as nickel–cadmium batteries) and that the obsolete inefficient battery is capable of using only 1%–2% of its original input energy.

8. The Relationship Between Technology and Society

I should like now to analyze the fundamental relationship between society and technology. Society has created technology as an object, and technology is the instrument for making products that enhance social life. In this sense, technology is dependent on society and has no self-standing. Society controls technology through the regulations that society imposes. If technology has a problem, then that problem is society's problem.

On the other hand, we are not likely to forget that technology changes society; in particular, since hand tools gave way to machine tools at the time of the industrial revolution, there was a change in means of production that led, finally, to the establishment of modern capitalism. The recent developments in automation and in computer technology have changed industrial society in profound ways. In this sense, technology is autonomous and develops of its own accord. Although technology cannot stand apart from the society that makes use of it, within its own domain technology is autonomous.

However, an examination of the relationship between technology and environmental problems reveals that strict environmental regulations have forced industry to develop technologies to cope with the problems it has created. Japanese regulations to prevent air pollution and the technology to treat wastewater are a result of the Japanese experience of *kogai*, a public nuisance. This is an example of how, through regulations, society can control technology, and the enactment of the regulations has been a breakthrough in society's efforts to tame the excesses of technological autonomy.

Although, on the other hand, a look at the autonomy of technological development reveals that smoke treatment and water treatment appear to be a development of "end-of-pipe technology", in fact their development at the moment is necessary as an essential step in the need for technology to become "clean". The reasonableness of a demand for cleaner technology has recently been stressed: Chapter 20 of *Agenda 21* (1992) insists on the need to ". . . prevent or minimize the generation of hazardous wastes as part of an approach to integrate cleaner production methods".[13]

One example of this has been a change in the method of producing sodium hydroxide from mercury process electrolysis to an ion-exchange method without mercury. We could evaluate this as an example of efforts to achieve cleaner production.

Recently, improvements have been made to fluidized-bed cement kilns and to the direct method or fused reduction method of iron and steel manufacturing. At the international level, in Germany, where environmental requirements are particularly strict, alternatives to CFCs are being actively developed and the recycling of vehicles is strongly promoted.

If, in the case of Japan, we reexamine the countermeasures that have been adopted, we find that Japan has taken a technology-oriented solution to the treatment of wastewater and gas. Yet, such strategies have their own limitations.

Although the treatment of gas has resulted in a reduction in the emission of sulfur oxides, there has been no similar clearance in the environmental standards of nitrogen oxides, whereas improvements to the diesel engine have been delayed due to technical difficulties. These are results of the inadequacies of the public transport system, the regulations regarding track transportation, and the workings of the social system itself. Because policies worked out with regard to technological improvements tend to be made before actions are taken to develop the technology, this may be too late as a policy response to necessary changes in the social system. If our aim is the effective conservation of the environment, it is important to ensure that there is a much better fit between social and technical demands.

References

1. Ministry of Health and Welfare, Japan (1993) A new comment on the Waste Disposal Law. Japan Environment and Hygiene Association, Tokyo
2. Washida T (1995) Market economy and resources recycling. In: Murota T (ed.) *Jyunkan no Keizaigaku* [Economy of circulation]. Gakuyo Shobo, Tokyo, p 160
3. Sirkin T, ten Houten M (1994) The cascade chain. Resources, conservation and recycling, Vol 10. Elsevier Science, New York, pp 213–277
4. Ueta K (1992) *Haikibutu to Recycling no Keizaigaku* [Economics of waste and recycling]. Yuhikaku, Tokyo, p 142
5. Yoshino T (1996) *Shigen Jyunkan Shakai no Keizai-Riron* [Economic theory of resources circulation society]. Tokai University Press, Tokyo, p 136
6. Environment Agency (1997) The white paper on the environment. Printing Bureau, Ministry of Finance, Tokyo, p 180
7. Pauri G (1996) Japan leads zero-emission [in Japanese]. Nihon Keizai Shimbun, August 24, 1996
8. Environment Agency, loc cit.
9. Schmidt-Bleek F (1994) *Wieviel Umwelt braucht der Mensch?* Birkhäuser, Basel
10. von Weizsäcker E, Lovins A, Lovins H (1997) Factor four, doubling wealth, halving resources use. Earthscan, London
11. Sachs W et al. (1998) Greening the north. Zed Books, London

12. Central Council of the Environment (1997) How to reduce the environmental burden created by waste disposal. Environment Agency, Tokyo
13. United Nations Conference on Environment and Development (1992) Agenda 21: programme of action for sustainable development. Rio Declaration on Environment and Development: statement of forest principle. United Nations, New York, Chapter 20

Chapter 1
Environmental Cleanup and Cost Bearing: The Methodological Problems*

1. Introduction

This chapter aims to analyze the method by which environmental cleanup and cost bearing are compared and assessed. One useful tool for the economic analysis of environmental problems is the theory of social cost. Yet, as I have pointed out in my book *Economics of the Environment and Technology*,[1] the traditional theory of social cost is flawed by theoretical confusion. The categories of social loss and social cost are not the same and should not be confused: whereas "social loss" is the loss of something's value to us, a real term category, "social cost" falls within the category of value terms.

The meaning of social cost includes the notion of irreversible loss, an absolute loss, and this is difficult to calculate as a value term. The category of social cost developed by William Kapp[2] included social loss itself, the first definition, and the prevention cost of social loss, the second definition. Because the second definition of prevention cost is not a social loss, Professor Shun'ich Teranishi noted the positive aspects of William Kapp's theory and proposed that the category of "social expenses" should be distinguished from the category of "social cost" as a means to make polluters pay and bear the cost of both prevention and compensation.[3]

An account of real environmental disturbances and damage from the standpoint of economic theory requires a three-level analysis: (1) an analysis of the real term value, value in use; (2) a categorical analysis of the value term; and (3) an analysis of the real expense in terms of money.

There are many levels of environmental damage. The first is the result of pollution at the regional level, as manifested in outbreaks of Minamata disease, Itai-Itai disease, and pollution of the air by a combination of petrochemicals at Yokkaichi. This kind of environmental damage includes not only damage to agricultural and fishery products, but also damage to human health. Because pollution in the air and water can easily cross national boundaries and can cause transboundary pollution, this type of damage becomes one constituent element in the catalogue of global environmental problems.

---

* Originally published in *Economic Journal of Hokkaido University* (Sapporo, 2000) 29:19–29. With the permission of the University.

The second type of environmental damage is that brought about by the progressive degradation of wetlands and forests, or pollution that leads to the weakening of bio-diversity and the natural ecosystem.

The third type of environmental damage is a lack of sunshine, of access to land-scapes, and the loss of natural amenities.

The overall problem that faces the global environment involves these three forms of environmental deprivation, as well as depletion of the ozone layer, disturbances to the global commons, and climatic change.[4,5]

At the level of problems caused by regional pollution, application of methodology based on an analysis of the real term value (A)—the theory of social loss, the real nature of the damage, the compensation awarded for the damage, restoration and reclamation of polluted land, and prevention of further damage—would be an analy-sis in terms of real value, value in use. An analysis based on categorical analysis of the value term (B), the theory of social loss from the standpoint of the theory of social cost, would evaluate damage in terms of trading value.

According to the degree of cost category (A), compensation and restoration work would be retrogressive and the cost would be negative, whereas (B) prevention work would be forward looking, seeking the prevention of future damage, and the cost would be positive. Should polluters find it difficult to bear the cost in terms of (A), the polluter would appeal for public funds, whereas in terms of (B), the polluters will try to internalize the expenses of prevention by combining them with rationalizing investment intended to reduce waste, and would, in part, take advantage of subsidies and tax privileges. It is an important feature of the situation that the negative and pos-itive costs should be closely related, because the incentives to pay the positive expenses designed to prevent future damage will depend on the setting of rules about how to allocate "negative cost".

In the third instance, an analysis of the real expense in terms of money, it is nec-essary to analyze the expenses related to environmental damage on the basis of the theories of social loss and social cost.

2. Social Expenses Incurred by Environmental Pollution Damage

Social expenses incurred as a result of damage caused by environmental pollution are basically intended to pay for compensation for damage to the environment and to pay for preventive measures against further damage, as well as for the restoration and reclamation of land damaged by pollution. These expenses will include the costs for negotiation and any lawsuits entered into.

2.1 Compensation for Damage

Compensation for damage covers damage to human health and life, and damage to property. The concrete matters covered by compensation for damage to health include direct compensation, cost of care, medical costs, and compensation for loss of revenue as a result of damage to agricultural or fishery products. In the case of agricultural soil pollution, compensation must be paid to those who had to cease planting because their land was poisoned. If a system to certify the existence of a pollution-related disease is set up, then victims can be identified and the realities of the damage these people have suffered can be made clear. However, this system may generate problems because it might overlook patients who are not officially registered and, thus, reduce the actual extent of the damage that has been caused.

2.2 Mitigation of Damage

The mitigation of damages refers to expenses paid out to mitigate or reduce the extent of real damage, as, for example, the provision of equipment near airports to reduce noise levels that are damaging to the health of local residents.

2.3 Restoration and Reclamation

Restoration and reclamation include the removal of cadmium from polluted soil, the removal of sludge contaminated by mercury, and the cleanup of contaminated soil using a chlorinated organic solvent. Should it be the case that damage to the environment is irreversible, measures to provide for alternative means of compensation are necessary. The task will involve not only the technical restoration of the contaminated area, but also its social restoration. Related issues will focus on "how clean is clean", and to what extent the polluters should bear the costs of the restoration.

2.4 Prevention

To prevent further pollution, it will be necessary to provide funds to pay for any machinery required to obviate pollution at the source. This machinery has been developed from "end-of-pipe technology", such as equipment designed to desulfurize fluid gas and apparatus to ensure that the production processes are themselves cleaner. Because it is possible to both conserve energy and material and to make a profit, the expenses for prevention cannot be counted as a social loss.

2.5 Transaction Costs and Administrative Costs

The transaction cost, namely the cost of negotiation and any ensuing lawsuits bearing on the costs for cleanup, as in the case of the USA Superfund system, nearly exceeds the cost of the cleanup itself. It is also necessary to provide funds for administering subsequent monitoring activities. The administrative cost depends on the system of cost allocation and the issue of "how to solve social conflicts".

The structure that provides for the generation of these particular expenses falls within the domain of the wider social system, and ". . . in order to reveal their origin, the study of social costs must always be an institutional analysis".[6]

3. Principles of Paying

3.1 The Pigovian Principle

The Pigovian tax is designed to tax the difference between the social cost and the private cost, or the difference between marginal private net product and the marginal social net product for the internalization of social cost, and it is not clear whether this would include the expenses listed in section 2 above.

A recent interpretation of the Pigovian tax states that where the cost curve of social marginal emission reduction and the cost curve of social marginal damage meet, each marginal emission that reduces cost is equalized,[7] and will include part of (2.1) the damage compensation, (2.4), and part of the expenses for prevention, which entails that this level cannot guarantee the prevention of absolute loss. Because the Pigovian tax aims to modify the price beforehand, it does not include expenses for restoration and reclamation.[8,9]

Coase's theorem has some bearing on the Pigovian Principle. Coase's theorem states that ". . . the delimitation of rights is an essential prelude to market transactions; but the ultimate result (which maximizes the value of production) is independent of legal decision".[10]

It therefore turns out that the minimization of social cost will be achieved independently of anyone's liability if the liability rule is invoked.

We must pay careful attention to the preconditions of Coase's theorem. Coase took Pigou's theory as a premise and set out to question it. Coase's main concern is not something's real value in use, but its trading value, and, for him, the object of economics is the maximization of social value: ". . . the question to be decided is: is the value of the fish lost greater than the value of the product while the contamination of the stream makes possible", and "It is all a question of weighing up the gains that would accrue from eliminating these harmful effects against the gains that accrue from allowing them to continue".[11] Coase's understanding is that because the missing value can be compensated for by the acquired value, the irreversible loss in real value is lost sight of from the very beginning.

Furthermore, Coase's theorem is suitable for application only if certain conditions are satisfied: namely, that the allocation of right is defined clearly, and that negotiations proceed without trouble, that is, with zero transaction costs. In the real world, this condition is not fulfilled.

Once the transaction cost is taken into account, then the decision about the distribution of resources depends on a rightful allocation. This is the latter part of Coase's theorem. Therefore, Coase's theorem focuses on both compensation in value and the transaction cost.

3.2 The OECD's Polluter-Pays Principle

The Polluter-Pays Principle (PPP) of the Organisation for Economic Co-operation and Development (OECD) is a well-known principle of payment levied on those who cause pollution. It states:

> The Polluter-Pays Principle means that the polluter should be charged with the cost of whatever pollution prevention and control measures are determined by the public authorities, whether preventive measures, restoration, or a combination of both. If a country decides that, above and beyond the costs of controlling pollution, the polluters should compensate the polluted for the damage which would result from residual pollution, this measure is not contrary to the Polluters-Pays Principle, but the Principle does not make this additional measure obligatory: in other words, the Polluter-Pays Principle is not in itself a principle intended to internalize fully the costs of pollution.[12]

The OECD's PPP is designated as a principle to ensure the rational allocation of resources and the correction of distortions in international trade, to assume charge of pollution control at the optimal pollution level, and, in exceptional cases, to compensate for subsequent damage; however, the OECD does not apply the principle to the general relief of damaged parties or to environmental restoration. Therefore, although the PPP does aim at (2.4) a certain level of prevention in very general terms, it does make exceptions for (2.1) compensation and (2.3) restoration. Beckerman interpreted the PPP as implying that ". . . the marginal damage to the victims cannot exceed how much it would cost them to avoid the damage, otherwise they would already have avoided it".[13] In fact, because savings were made in the expenses for prevention, huge damage has resulted.

It is important to our understanding of the PPP that, although it first of all indicates that the original polluters must pay, it does not concern itself with the possibility that payment may be shifted from the polluter to the consumer, or that the polluter may take measures to internalize the cost. Therefore, we should not overestimate the power or effectiveness of the PPP.

3.3 The Principle of Paying Capability

The principle of paying capability has to be in place because the responsible party may, in an emergency, be incapable of paying, as, for example, in the case of Japanese aid in the dismantling of old Soviet nuclear submarines. However, if such a case drags out for a long time, it will contradict the policy of social justice and there will be no incentives for the responsible party to reduce pollution.

3.4 Benefit Principle

The benefit principle is a best fit for expenses incurred as "positive expenses for the future". That is, the beneficiaries of environmental conservation must combine to pay the expenses of positive environmental conservation.

3.5 The Principle of the Responsible Party

The principle of the potential responsible party is designed to operate in cases of insolvency or where the polluter is unknown and must be sought for, as is the practice with the USA Superfund system. Therefore, this principle will, if the priority between the interested parties is not clarified, lead to lawsuits, increase transaction costs, and reduce economic efficiency. To avoid this, both The Netherlands and Germany have passed legislation to clarify the ordering and priority of the interested party. At the same time, if a wide definition is given to the extent of the responsibility of the potential responsible party, there will be greater incentive to reduce pollution and greater encouragement to carry out intensive investigation (thus increasing the transaction and investigation costs).

4. The Cost-Bearing Rule

Who, finally, has to bear the cost? Different social systems have different means of responding to environmental damage. We can refer to three different examples: (1) the Japanese polluter-pays (bearing) principle; (2) the USA Superfund system; and (3) the Dutch and German soil protection acts. In addition, there is the principle of joint compensation and help from the public purse. The criteria for assessing the cost-bearing rule are: (1) how far the rule reduces pollution and improves the environment; (2) how it accords with social justice and fairness; and (3) how much it serves to bolster long-term economic efficiency.

4.1 The Japanese Polluter-Pays (Bearing) Principle

While the OECD's PPP is not in itself a principle intended to internalize fully the costs of pollution, the Japanese Polluter-Pays (Bearing) Principle has extended the interpretation of the OECD's PPP to take on the character of liability law, and includes the cost of environmental restoration within the cost of accumulated pollution and damage relief.

The 1975 Japanese *White Paper on the Environment*, edited by the Environment Agency, remarks (Chapter 3) that:

> The cost for environmental conservation includes the cost of pollution control, the cost of environment restoration, the cost of damage relief, the cost of avoiding pollution and the administrative cost. While the OECD's PPP is mainly aimed to prevent the distortion of international trade and the cost of pollution control, the Japanese PPP includes the costs for environment restoration and damage relief.[14]

The Japanese PPP is incorporated into the 1973 Pollution-Related Health Damage Compensation Law and the 1970 Law Concerning the Entrepreneur's Bearing of the Cost of Public Pollution Control Work. Consequently, the Japanese PPP includes: (1) damage compensation; (2) damage mitigation; (3) restoration and reclamation; (4) prevention; and, at the same time, (5) the administrative cost, which includes the cost of monitoring, that must be borne by the Environment Agency and local government authorities. Although the extent of the designation of the Japanese PPP is apparently rather wide, neither the Pollution-Related Health Damage Compensation Law nor the Law Concerning the Entrepreneur's Bearing of the Cost of Public Pollution Control Work actually applies the principle strictly. There are no formal regulations and there is no law dealing with soil contamination other than that stipulated for contaminated agricultural land.

4.2 The USA CERCLA

Although it lays down no regulations for restoring environmentally damaged land, the USA 1980 Comprehensive Environmental Response, Compensation and Liability Act (CERCLA) does designate liability for the cleanup of hazardous substances, and, in cases where the cleanup is carried out by the Environmental Protection Agency (EPA), the costs are paid out of the superfund. Moreover, the responsibility is retroactive, and covers joint and strict liability, the liability of present and past owners, and the lender's liability.

Because this law focuses mainly on the responsibility for the cleanup and for the allocation of costs, there have been many lawsuits between potential responsible parties and the EPA, with the result that the costs of transaction are greatly increased and the cleanup is delayed. On the other hand, because companies know that they will be asked to spend large sums of money for compensation and cleanup, they have become more cautious, and now take more steps to prevent pollution in the first place.

4.3 The Soil Protection Acts of The Netherlands and Germany

The 1994 Dutch Soil Protection Act[15] aims to reduce the threat to the functional properties of the soil by designating the disposal of waste to prevent soil contamination and by empowering the state to recover the costs of remediation from the polluter or the owner. The owner or long-term leaseholder of a property is not responsible in cases where the owner: (1) has had no sustainable legal relationship with the polluter or polluters during the period in which the pollution occurred; (2) has had no direct or indirect involvement in the cause of the contamination; and (3) was not aware, or, in all fairness, could not have been aware, of the contamination at the moment of acquiring the title of the property. Therefore, the aim of the Dutch system is to investigate and restore the environment on the premise of preventing contamination.

The German Soil Protection Act of 1998 is in line with the Dutch model.[16]

4.4 Joint Compensation System

The joint compensation system has been recommended in a Green Paper produced by the European Commission (EC).[17] In cases where the responsible party is not specified or the extent of the pollution is not clear, civil liability is not an effective tool. The joint compensation system is an application of the PPP if it is supported by an industrial group that has the capability of causing environmental damage. The joint bearing is the principle that liaises between private bearing and public bearing. The International Oil Pollution Compensation Fund is a sort of joint compensation venture, and, because the fund already exists, it is possible to respond to emergencies and provide immediate cleanup, even in cases where the individual cannot pay to provide compensation. However, problems remain: who has the right to authorize use of the fund, who is to manage it, how clean is clean, and (because of cost dispersion) an increasing lack of incentive. The question of fund raising was reexamined at the time of the revision of the Japanese Waste Disposal and Public Cleansing Law to determine the restoration of an illegally used dumping site.

At that time, meetings of the Ministry of Health and Welfare Living Environment Commission debated the issue of cost bearing with some heat. In the end, the revised act included the new item that ". . . businesses are required to fund at the center for promoting appropriate treatment". The Japanese Federation of Economic Organization (*Keidanren*) asserts, against this item, that where business is carrying out appropriate treatment, it should not bear the cost of restorations. In addition, when an alternative to cost bearing by businesses without certification of appropriate treatment was proposed, business saw that this would impose the costs of certification on it and strongly opposed the proposal. It seems that business feared that commitment to voluntary funding would involve it in heavy expenses when the cleanup was of a serious nature.

Under the system of the USA superfund and Resource Conservation and Recovery Act (RCRA), even the cleanup is paid for by the fund, whereas the polluter is merely reprimanded for introducing a "moral hazard". Yet, because paying for the cleanup out of public finds weakens incentives to institute good environmental management, such a course of action itself tends to induce "moral hazard". Similarly, the use of the safety net of public funding to solve bad bank loans does not necessarily lead to the healthy management of a bank.

The use of public funds to manage environmental conservation is based on notions of community and cooperativeness. As today's environmental infrastructure grows wider and wider, it becomes more and more necessary for it to examine itself. The environmental infrastructure is constructed as a public business devoted to such environmental duties of reclamation as cleanup. The Law Concerning the Entrepreneur's Bearing of the Cost of Public Pollution Control Work uses such work as a reason to reduce the polluter's bearing with phrases like ". . . functions other than pollution control".

5. Conclusion

Finally, we have to address the responsibility of the state for environmental damage, and its consequent liability. If the state itself, or an enterprise in public ownership, causes pollution, then the state is the direct polluter and is clearly responsible.

In cases such as the outbreak of Minamata disease, the state can be held liable, because although it is not the direct polluter, it has admitted the occurrence of pollution while ignoring the plight of the victims and, at times, even treating the victims as offenders. How should we theorize or construct the responsibility of the state? In purely juristic terms, the main problems are the relevant clauses and the enforcement of the act dealing with water pollution. It might be possible to make a quasi-application of the PPP to include the state's responsibility and to theorize the state's support for the Chisso company, the company directly responsible for the pollution. Or, it might be possible to include the state and the local government among those parties who are potentially responsible. This may apply to the case of Teshima Island, where illegal waste was dumped with the connivance of the local authorities.

At any rate, the problem is large and its solution is difficult. Although I do not defend any easy reliance on the financial support of the state, the responsibility of the state must be pursued because, if it is not, then the state's administration and standards of environmental control will become very lax.

References

1. Yoshida F (1980) Economics of the environment and technology. Aoki Shoten, Tokyo
2. Kapp W (1950) The social costs of private enterprise. Harvard University Press, Cambridge MA
3. Teranishi S (1984) The problems of social loss and the theory of social cost. Hitotsubashi Ronso, 91:22–41
4. Teranishi S (1992) *Chikyu Kankyo Mondai No Seiji-keizaigaku* [Political economy of global environmental problems]. Toyo Keizai Shinposha, Tokyo, Chapter 5
5. Daly H, Cobb J Jr (1989) For the common good. Beacon, Boston, Appendix

6. Kapp W (1963) Social costs and social benefits—a contribution to normative economics. In: Beckerath EV, Giersch H (eds) *Probleme der normativen Ökonomik und der wirtschaftlichen Beratung*. Duncker & Humbolt, Berlin, p 186
7. Ueta K (1996) *Kankyo Keizaigaku* [Environmental economics]. Iwanami Shoten, Tokyo, p 119
8. Andersen M (1994) Governance by green taxes, making pollution prevention pay. Manchester University Press, Manchester, pp 44–45; Anderson pointed out that the purpose of externality taxation is not to compensate, but to impose restraints on polluters
9. Bowers J (1997) Sustainability and environmental economics, an alternative text. Longman, Harlow, Chapter 5; this is a critique of the Pigovian model of pollution
10. Coase R (1959) The Federal Communications Commission. J Law Econ, II:27
11. Coase R (1960) The problem of social cost. J Law Econ, III:2, 26
12. OECD (1975) The polluter-pays principle. OECD, Paris, p 6
13. Beckerman W (1975) The polluter-pays principle: interpretation and principles of application. In: OECD (ed.) The polluter-pays principle. OECD, Paris, p 50; the background to the introduction of the PPP in the USA seems to have been problems of "pollution dumping"
14. Environment Agency (1975) White paper on environment. Printing Bureau, Ministry of Finance, Tokyo, p 74, and Amano A (1997) *Chikyu Onndan-ka no Keizaigaku* [Economics of global warming]. Nihon Keizai Shimbun, Tokyo, pp 31–33; Professor Akihiro Amano criticized the Japanese Polluter-Pays (Bearing) Principle as a "... backward legal principle" compared with the OECD's PPP. I agree with Professor Amano that subsidy to the polluter goes against the OECD's PPP, but the cleanup of accumulated pollution and compensation has become a common principle in Europe and the USA as having preventative effects
15. Ministry of Housing, Spatial Planning, and the Environment (1994) Soil protection act 1994. The Hague
16. "Gesetzes zum Schutz des Bodens" vom 17 März, 1998. Artikel 1 Gesetz zum Schutz vor schädlichen, Boden veräderungen und zur Sanierung von Altlasten
17. European Commission (1993) Green paper on remedying environmental damage

Chapter 2
Environment Politics in Japan*

1. Introduction

> "Development can be seen as a process of expanding the real freedom that people
> enjoy."
> Amartya Sen[1]

Amartya Sen argues that the growth of the gross national product (GNP) or of
individual incomes is very important as a means of expanding freedom, but freedom
also depends on social and economic arrangements, and on political and civil rights.
Sen lists five distinct types of "instrumental freedom": (1) political freedom; (2) economic facilities; (3) social opportunities; (4) transparency guarantees; and
(5) protective security.

In the light of Sen's analysis, I should like to offer some considerations on the nature
of the results (or disintegration) of Japan's development and environmental policies.
We can consider these under three headings: (1) the comprehensiveness of the development objectives; (2) the policy style and its democratic implications; and (3) the
application of development objectives as a policy tool.

1.1 *Comprehensiveness of Development Objectives*

Since the 1960s, Japan's economic policy has, in accordance with the famous
income-doubling policy, been motivated mainly by setting targets for the growth of
income. Consequently, as an income-doubling target was achieved and the income differences were reduced, considerations such as the quality of life, culture, amenities,
leisure facilities, and opportunities have been given a relatively low priority. In
response to the negative effects of development on the environment and the health of
the citizens, the policy adopted for issues of health and safety has been to "react and
cure" (rather than prevent). A Japanese saying sums up the situation nicely: "Only

---

*This paper was originally prepared for the project on Japanese Environmental Policy
organized by the World Bank Institute under Brain Trust Program. With the permission of the
World Bank Institute.

gai-atsu (pressure from abroad) and human tragedy can bring about changes in Japanese policy".

As a result, those issues that affect the environment or reduce amenities, but do not harm human health directly, have again been given low priority.

1.2 Policy Style and Democracy

After World War II, the Japanese political and economic systems were obliged to become more democratic, while, at the regional level, local governments were set up, and the practice of electing mayors and prefectural governors was inaugurated. Ever since, local governments, supported by citizens movements, have had a more important part to play in the implementation of strict environmental regulations than national agencies.

Meanwhile, the central government has retained its old bureaucratic habits and has changed only partly, while there has been little lessening of the factionalism (as well as sectionalism) of ministries and agencies, thus leading to a serious lack of coordination in overall governmental policy.

One result of this has been that the planning processes of schemes for development have not been open either to the general public or to the Diet, where the schemes have not featured on the agenda for discussion. Therefore, there has been little public debate, formal review, or scrutiny of the development plans. Nor have citizens who are likely to be affected by a project been permitted to take part in the real planning process.

1.3 Policy Tool for Development

One of the economic tools used for development projects, particularly public works schemes, has been government subsidies. In comparison with other countries, Japan's percentage rate of general government fixed capital formation in terms of GNP is two- to threefold higher. This means that government expenditure on public works projects has a wide-ranging and deep effect on the economy.

One side effect of this is the standardization and uniformity of urban planning and road construction at the local level, which tends to overlook regional amenities and demolishes traditional landscapes.

In addition, while on the one hand publicly subsidized works lack flexibility and tend to enlarge the size of the budget, government subsidies can, on the other hand, lead to the shortening of the life span of the facilities. It will have become obvious by now that there is also a strong link between the conduct of public works projects and political power. Critics point that because many local councilors are members of construction-related industries, we have a situation that is referred to as "grass-route conservatism" (this is not a misspelling, but a typical Japanese verbal joke, taken from the English term "grass-roots democracy", and meant to be a pun). This is a feature of what is known as *Doken-Kokka*—the Construction State (another somewhat cynical joke). Inevitably, the main interest of members of local councils will be how to get hold of and spend the subsidies allocated to their budgets.[2]

Focusing on the above-mentioned three issues, this chapter analyses: (1) national land/regional development; and (2) energy policy and the environment.

2. National Land/Regional Development and the Environment

"Agencies and ministries seem to act independently in a spirit of competition rather than working together in a fully co-operative way."   OECD[3]

Japan has a relatively bad record when it comes to the integration of its development and environmental policies. Indeed, Japan has made many mistakes in this area. In many cases, plans for development have included only a small consideration of measures designed to protect the environment as a central feature of the plans. Consequently, development projects have often severely disturbed the environment. Many of these negative effects can be traced back to Japan's "bubble economy". It is therefore essential that we should analyze these failures and make a clear and objective assessment of their causes.

The most negative aspect of Japan's post-war development policy has been the harmful environmental effect of the ill-considered National Consolidated Development Plan.

Soon after the end of the war, the government designated a number of locations as part of a regional development plan (*Tokutei chiiki kaihatsu*) aimed at utilizing water to provide hydroelectric power. The construction of dams and reservoirs resulted in the flooding of considerable areas of valuable rural land.

2.1 New National Consolidated Development Plan (1962–)

In 1962, the Government launched a New Consolidated Development Plan based on the Growth Pole Strategy (*Kyoten Kaihatsu Shugi*). This envisaged a string of large-scale petrochemical plants and as well as iron and steel manufacturing complexes along Japan's Pacific coast, an area that would embrace the localities of Yokkaichi, Chiba, and Sakai. The plan proposed the creation of 15 new industrial cities and six development areas.

The results, however, have not been benign, and, among the harmful consequences, we can note:

1. The disruption and pollution of the environment in the vicinity of these complexes;
2. A deterioration in the quality of neighborhood agriculture and a weakening of small businesses;
3. A crisis in the budgets of the local governments; and
4. A weakening in the autonomy of local governments as a result of their dependence on the national subsidy.

As causes of these negative effects, we conclude:

1. That capital investment and the overhead social costs were mainly devoted to creating the industrial infrastructure, and much less attention was paid to the prevention of pollution;
2. That the local governments reaped little tax revenue because they offered tax reduction policies in order to attract industrial complexes to their localities;

3. That the budget for providing social welfare was reduced to pay for pollution prevention measures; and
4. That there was a delay to provide facilities for desulfurization, for low sulfur oil, for high stacks, and a green belt in and around the plants. A typical case involved the problem of pollution at Yokkaichi Industrial Complex, which finally came up before the court.

Critics have argued that the New Consolidated Development Plan was a failure because it was based on short-term economic targets and was never intended to enhance or enrich the quality of people's lives, their culture, education, and health.

2.2 The Second National Consolidated Development Plan (1969-)

It was in this expansionist climate that the Second National Consolidated Development Plan was launched in 1969. It proposed to build a gigantic industrial complex, distributed throughout Japan, aimed at the full realization of national land potential, but with little reference to preserving a harmonious balance between humans and nature. Immense projects were set in motion, like those at Mutsu-Kogawara (Aomori Prefecture), east Tomakomai (Hokkaido) and Shibushi (Kagoshima Prefecture), backed up by interest groups and parliamentary politics. The government of Prime Minister Kakuei Tanaka initiated a campaign called The Plan for Remodeling the Japanese Archipelago (*Nippon Retto Kaizo Ron*, 1972),[4] which was designed to stimulate the construction of roads and infrastructure throughout Japan; unfortunately, because strict regulations concerning land policy were not enforced, the plan also encouraged private companies to speculate in land investment.

At the same time, plans for the establishment of a national network system to link everywhere in Japan with bullet trains were also drawn up.

There were a number of gray areas in these plans that led inevitably to problems and severe disputes, including: (1) uncertainty over the object of development; (2) uncertainty over the agency responsible for the development (the third sector); and (3) uncertainty as to who should be held responsible for the results of the development.

In response to critics of the measures, the person in charge of the National Consolidated Plan (Deputy Secretary of the National Land Agency at that time) said, "Development policy is a process of trial and error".[5] The question, however, remains: who is to be held responsible for the harm caused by such error-prone hit-and-miss development policies, and how should such persons be held accountable?

Because the plan for a huge industrial complex had overestimated the demand for products (iron, steel, and petrochemical goods), and as a result of further changes in the industrial structure after the oil crises of 1973 and 1979, the whole plan lost its feasibility. Yet, although this was clearly going to cause serious problems, fundamental reconsideration of the original plans was postponed.

In the cases of east Tomakomai and Mutsu-Kogawara, for instance, the original plan was scaled down, but it was not changed fundamentally. Consequently, large areas of land that had been bought up for industrial development have remained unused and it has not been possible to recover the mountain of bad debts incurred in purchasing the land; repayment of the debts fell to the tax payer.

Case Study: The Failure of the Mutsu-Kogawa Development Project[6]

The project for the Mutsu-Kogawara gigantic industrial complex was launched in 1969 as a part of the Second National Consolidated Plan. *The Report by Japan Industrial Location Centers*, published in 1969,[7] proposed the building of an enormous complex that would include facilities for the production of iron, steel, petrochemical goods, and nuclear power. The original proposal was premised on the supposition that there would be huge demand (both at home and abroad) for such goods. In addition, the original proposal omitted to take into consideration a number of vital issues, including:

1. That the plan treated the region as if it was an unoccupied space and ignored the fact that people lived in the area;
2. The plan paid little attention to the local industries of agriculture and fishing;
3. The plan showed no consciousness of the need for the parallel development of industry and agriculture; and
4. No steps were taken (or even suggested) to prevent unregulated land speculation.

A dual system of development was implemented. While Aomori Prefecture set up a public corporation to manage the development, an agency of the central Government, *Keidanren* (the Federation of Economic Organizations), and Hokkaido-Tohoku Development Finance Corporation established a special committee. Consequently, no one was sure who actually had overall responsibility for implementing the plan. At the same time, just after the plan was made public, there was a flurry of large-scale and unregulated land speculation, carried on without regard for the context of industrial investment.

In 1973, during the first oil crisis, a local people's protest movement began to gather force; yet, even after 1974, when the economic situation clearly worsened, speculators continued to buy up the land, which only ensured that the debts incurred by the Public Development Corporation grew bigger and bigger. At the same time, the farmers who had sold their land lost all chances of reemployment because the proposed industrial plants remained unbuilt: such farmers became known as the "refugees of development" (*Kaihatsu-nanmin*).

The major defects of the whole process were:

1. The very short term allowed for the decision-making process (13 months);
2. The autocratic and unilateral behavior of the Aomori Prefecture Government, which had no intention of listening to what the local people might want to say;
3. The vagueness of purpose in the department of the central government responsible for the process;
4. The inflexibility of the Aomori Prefecture Government and its inability to reconsider the project, as well the central government's ratification of the project; and
5. The handing over of the plan to a party with little real political power.

During the 1970s, only the plan to build oil storage plants was actually realized, and, by 1983, the amount of the loan had reached 139 billion yen. In 1984, a new national project to build plants designed to reprocess nuclear fuel and to process uranium was launched, and, in 1985, the two responsible bodies—the Federation of Electric Power Companies and the Public Development Corporation—reached an agreement over the location of these facilities. However, because no decisions were taken about the location of other projected facilities, the amount of bad debts reached 140 billion yen, consisting of money that had been loaned for the purchase of land from farmers, and to compensate for losses sustained by the fishing industry.

Public opinion in Aomori was divided between those for and against the installation of nuclear facilities, usually according to the degree of political conservatism (those for) or the opposite (those against). A subsidy of 19 billion yen had already been provided to Rokkasho Village for the location of nuclear-related facilities.

The failures of the whole process stem from:

1. The overextensive reach of the petrochemical complex project;
2. The lowering of the basic living conditions of people domiciled within the development area, and the serious breakup of social relationships between local communities;
3. The uncritical acceptance of "unfavorable" facilities: the invitation stage of the development (1st stage); the dependence on development (2nd stage); and the acceptance of dangerous facilities (3rd stage); and
4. The accumulation of bad debts incurred by the Public Development Corporation (the Corporation went bankrupt, with bad debts of 240 billion yen, in 1999).

The problems to be scrutinized are:

1. The immense scale of the development project;
2. The dependence of the success of the development on changing conditions in the outside world;
3. The dependence of the development on special conditions;
4. The plurality of agencies responsible for the development; and
5. The failure to estimate likely costs.

2.3 The Third National Consolidated Development Plan (1977)

In 1977, the Third National Consolidated Development Plan was enacted. In 1983, the Technopolis Law (the Law for Promoting the Development of High Technology Integrated Regions) to promote the location of high-technology industry in designated areas was introduced. Twenty-six technopolises were designated (as of 1988). The leading industries were to be electronics, semiconductors, optics, and other lightweight and high value-added products, as well as facilities for research and development (e.g., computer software). Inland development was supported by the transportation systems (highways, airports). Smokestack emissions were reduced, but high-technology pollution problems have arisen.[8]

2.4 The Fourth National Consolidated Development Plan and Resort Development (1987–)

Ever since the central Government's attempt to implement the New National Consolidated Development Plan of 1962, it has also been promoting schemes to encourage tourism, but it was not until the 1987 Fourth National Consolidated Development Plan and Resort Law was passed that tourism became the target of large-scale development throughout Japan as part of the Government's plan to revitalize the private sector and to vitalize excess liquidity of money among private businesses after pressure from the USA to enlarge the domestic market following the high valuation of the yen after 1985.

The wording of the Resort Law makes its premise and purpose quite clear: Item 1, to develop resorts in areas of well-preserved natural amenities by encouraging private company initiatives; Item 14, to permit the radical deregulation on farmland for diversion for other purposes; Item 15, to open the national forests as resort areas; and to provide subsidies and tax reduction to encourage resort development.

As a result, because all 47 prefectures immediately initiated resort projects (by 1990, there were 900 such ventures), approximately 20% of all Japan's total landmass has been earmarked as an area for resort development.

This movement stimulated land speculation and a sharp rise in the price of land, and was one of the causes of the bubble economy of the late 1980s and early 1990s. Thus, the Resort Law became one of the pillars of the bubble economy and a source of bad investment by providing the scheme outlined below.

In order to qualify for the central government's subsidies and tax reductions specified by this law, local governments drew up plans to acquire the permission of landowners to develop land while inviting companies to invest in the development of the land. The private companies borrowed money from banks in order to invest it in the construction of golf courses, ski resorts, and holiday hotels. Because the project was administered as a third-sector system in which local governments and private developers were jointly involved, the development of private resorts was, in part, treated as an issue of public works.

At the same time, the responsibility and control for each third-sector project tended to be vague. Therefore, the combination of the National Consolidated Development Plan and the Resort Law was the main cause of rampant land speculation and the immediate cause of the bubble economy.

Now, after the bursting of the bubble, we are left with widespread environmental disturbance and extensive areas of land abandoned after the withdrawal of developers, while the burden of paying off the bad debts falls to the tax payers.

We are obliged to point out that the originators of the initial plan that led to this mess are short-sighted lawmakers, and it is they who should be held responsible.[9]

2.5 Summary: The Nature of Failure

We may sum up the problems related to the environmental disturbances caused by the development plan as follows:

1. The secrecy of the decision process, which was open neither to the general public nor even to the Diet; only the Cabinet discussed and acknowledged the matter;

2. The absence of any official review and scrutiny process for the plan, which tended to overestimate the demand for industrial goods and overlooked the changes that were occurring in the structure of industrial undertakings;
3. The initial failure to designate a responsible ministry or agency to implement the plan, which, consequently, led to a failure of accountability and responsibility (although, after 1974, the Government did set up the National Land Agency);
4. The encouragement that the announcement of the plan offered speculators to invest too hastily in land development;
5. The lack of strict regulation of land speculation and the acquisition of private land, which was an essential condition in stimulating overdevelopment; and
6. A failure to include as central parts of the plan any measures designed to protect the environment.

Case Study: Domestic Development as an Alternative

In contrast with the traditional development pattern that has depended on external forces directed by national interests, many regions of Japan have eagerly sought alternative means of regional development. This style of self-directed development, known as domestic development, includes a number of distinguishing features,[10] as outlined below.

1. Domestic development is not dependent on subsidies from the central Government or on invitations to big companies, but bases its operations on its own regional resources, industries, and culture, so that the people of each area will learn to develop and manage their enterprises themselves;
2. By paying attention to environmental preservation and having due regard for each region's nature, landscape, and natural amenities, domestic development seeks the sustainable development of local societies;
3. Domestic development seeks to construct a complex industrial structure with as close and as multiple a relationship with the region as possible, thereby preventing a leakage of profits generated within that area and so ensuring a high-quality economic structure; and
4. Such a pattern of regional development institutionalizes the involvement of citizens in a grass-roots democracy.

We can find many examples of domestic development throughout Japan, including: (1) Oh-ita Prefecture's initiative in sponsoring the movement to foster the idea of "one village, one specialty"; (2) Kanazawa City's plan to preserve its historical heritage by encouraging the furtherance of traditional industries and crafts; and (3) the creation of Ikeda Town's (Hokkaido) own winery.

At the same time, plans have been laid down to stimulate programs for the creative use of Japan's mountainous regions.

Activation programs take various forms, as outlined below.

1. Type A: the support of urban residents for people living in mountainous regions;
2. Type B: the trade and exchange of agricultural and forest products; for example, the direct sale of local goods to urban areas, and the ownership of orchards by people living in urban areas (use of economic instrument);

3. Type C: investment in forests, houses, or land by urban dwellers; for example, by the purchase of second homes (use of economic instruments);
4. Type D: mutual help between communities located along the same river; for example, forestation paid for out of reservoir funds;
5. Type E: cooperation between urban and rural areas to develop resort or health villages; and
6. Type F: self-management of villages through organizations such as agricultural cooperatives.[11]

3. Energy Policy and the Environment

3.1 Subsidy for the Location of Power Stations

Japan's energy policy after World War II had a deep impact on the environment— from hydroelectric power-generating dams, and coal- and oil-powered generating stations to stations generating nuclear power. Each type of generating station has many environmental effects and has caused many problems.

Here, I would like, in particular, to analyze the three laws passed in 1974 to promote the development of electrical power resources, namely the Law for the Promotion of Electric Power-Resources, the Special Account Law for Promoting the Development of Electric Power Resources, and the Law to Manage the Area Surrounding the Electric Power Facilities. These laws were introduced to promote the development of electrical power stations against the background of movements opposed to the construction of power stations, and nuclear power stations in particular. I undertake this analysis here because these three laws have recently been attracting the attention of other Asian countries.

In 1973, the government of Kakuei Tanaka, which had already introduced an automobile fuel tax to pay for road construction, introduced these three laws to promote the construction of power-generating stations by using a tax levied on the consumption of electricity. A subsidy drawn from the Power-Resources Development Levy, partly intended to protect the environment, was used mainly for public works projects, such as the construction of roads leading to and from the power stations.

A pamphlet issued by the Resources-Energy Agency explains the purpose of the three laws in these terms:

"Since Power Generating Stations employ fewer people than do other industries, and because the energy is not utilized at the sites near the stations but is relayed to metropolitan areas, we have to be able to answer the complaints of local people who see no merit or gain to themselves in siting the stations in their locality".[12]

Therefore, we can see that this system amounts to a form of bribery used by the devisers of the scheme to pacify users of the system who might otherwise oppose it. For example, during the years 1974–1984, the total subsidy granted to this system amounted to 266 billion yen, yet, in the early 1980s, an individual user paid 1000 yen per year as a resources tax. Recently, although 300 billion yen has been allotted as a subsidy to this system each year, over 100 billion yen is surplus to expenditure. The subsidy is apportioned between educational and cultural facilities (29%), road construction (25%), sports and recreational facilities (15%), and agriculture and fishery facilities (8%). The prefectures that received the largest proportion of the subsidy were those in which nuclear facilities had been sited, namely, Fukushima Prefecture and Fukui Prefecture.

It has been pointed out that this system caused several problems. As the subsidy swells the temporary budget, the cities, towns, and villages that receive it tend to build unnecessary facilities, which need expanded maintenance expenditure for their upkeep. At the same time, the fixed assets tax derived from power-generating facilities decreases over time. Nevertheless, the cities, towns, and villages in which the facilities are located tend to accept the additional facilities.[13]

Consequently, Japanese nuclear power-generating stations are concentrated locally in Fukushima Prefecture, Fukui Prefecture, and Niigata Prefecture, one in each prefecture.

At the same time, this system entails that much of the energy generated is not consumed locally, where the population density is thin, and so it is not used for cogeneration (for heating), whereas people in metropolitan areas have no idea where their energy is coming from, and often use it wastefully. This means that a good deal of energy is not used efficiently, and people do not bother to weigh up the risks and benefits of using nuclear energy in daily life.

3.2 Nuclear Related-Facilities and the Environment

The other big issue raised by the energy policy is the effect of nuclear energy on the environment. Japan is home to 51 nuclear power plants and related facilities, and already one-third of the electricity used in Japan is generated by nuclear power stations. The management and safety of the nuclear power stations is not regulated by the Environment Agency, but by the Science and Technology Agency and the Resources Energy Agency of the Ministry of International Trade and Industry (MITI), and, ever since the early 1970s when commercial nuclear power stations began to come into operation, many small-scale accidents have occurred at nuclear power plants, for instance, leaks of steam, and cracks in the feed water pumps due to metal fatigue. Recently, however, more serious accidents have occurred, including in 1997 at the experimental fast breeder reactor (FBR; Monju) operated by the Power Reactor and Nuclear Fuel Development Corporation (Fukui Prefecture), and in 1999 at the conversion plant (JCO) at Tokai-Village (Ibaraki Prefecture).

Because Japanese nuclear policy aims to be able to manage the complete reprocessing and nuclear fuel cycle, Japan has built many sorts of facilities capable of reprocessing and maintaining a nuclear fuel cycle. The small-scale plant at Tokai-Village is a unit in the reprocessing plan, whereas a larger reprocessing plant is now under construction at Rokkasho-Village (Mutsu-Kogawara), although much spent fuel is still being transported to and reprocessed in the UK or France. Although the government proposes to build a FBR that will be able to use plutonium, the major accident at Monju indicates that the successful (and safe) development of a FBR may remain elusive. Nevertheless, the reprocessed plutonium that has been returned from the UK and

France is being stockpiled, and the government is planning to use plutonium in a light water-type reactor (plu-thermal).

As for safety regulations, a nuclear safety committee is attached to the Science and Technology Agency, but there are only approximately 30 staff, while responsibility for ensuring that safety regulations are observed also falls within the remit of the Science and Technology Agency and the Resource Energy Agency. This means that both the responsibility for the safety of the operation and the direction of the operation itself falls to the same authority. This differs from the policy in the USA, where the agencies for regulation and development (the Nuclear Regulation Committee; NRC) are kept separate.

With regard to nuclear waste, no formal law exists on how to store and deal with high-level radioactive waste, nor is any guidance offered on how to select candidate deposit sites of high-level radioactive waste. At Rokkasho-Village, which is still under construction, high-level radioactive waste is already being stored in glass canisters.

4. Conclusion

In the Introduction, I proposed three analytical points of views for focusing on Japanese environmental politics. Let me summarize below.

4.1 Comprehensiveness of Development Objectives

Recently, the long-term environmental effects of global warming, or of substances that disturb the human endocrine system, have become major items on the environmental policy agenda. At the same time, the Environmental Impact Assessment (EIA) process has been introduced at the national level to deal with problems arising from industrial development, including major construction projects or the building of power plants.

Nevertheless, we need a more comprehensive overall social assessment of public works, because many of these public projects have a negative effect on the environment, as well as ignoring dictatorially the democratically expressed will of the people, not to mention incurring a budget deficit. Furthermore, in terms of sustainable development in the long run, we need an "environmental new deal" (public works for environmental restoration) to deal with the cleanup of contaminated soil in metropolitan and rural areas, and for the removal of traces of heavy metal, organic solvents, and other toxic substances.

4.2 Policy Style and Democracy

In many cases, not only do the plans tend to overestimate the future demand for new facilities but, because the plans have been unable to forecast the actual consequences of the structural changes, the plans have also been unable to make adjustments accordingly. There have been no formal assessments of the final results and no reconsideration of the plan once it has been given the go-ahead.

Consequently, it is not clear who bears responsibility for the implementation of the development projects or who should be held accountable when the projects go wrong or prove harmful. This problem highlights the issue of transparency in the workings of Japanese democracy.

4.3 Policy Tool for Development

It is clear that few things could, therefore, be more important than a reform of the subsidy system. With regard to subsidies that are granted to prevent pollution, both central and local governments operate many kinds of system, whereas, as in the case of Itai-Itai disease, tax reductions and special depreciation systems have been made available to deal with pollution. These systems run partly counter to the PPP of the OECD.

At the same time, because the Japanese subsidy system focuses mainly on the facilities themselves, environmental policy tends to be technology orientated, rather than directed toward the making of necessary changes in the social system. However, to protect the environment we need changes in government attitudes toward the needs and well-being of society as well as to the operations of technology.

References

1. Sen A (1999) Development as freedom. Alfred A. Knopf, New York
2. Miyamoto K (1973) *Chiiki-Kaihatu ha korede yoinoka* [A critic to regional development], Iwanami Shoten, Tokyo
3. OECD (1994) Environmental performance review of Japan. OECD, Paris, p 113
4. Tanaka K (1972) *Nippon Retto Kaizo Ron* [The plan for remodeling the Japanese archipelago]. Nikkan Kogyo Shimbun, Tokyo
5. Honma Y (1991) *Kokudo Kaihatu keikakuno Shiso* [Idea of consolidated national plan]. Nihon Keizai Hyoronsha, Tokyo
6. Funahashi H, Hasegawa K, Iijima N (eds) (1998) *Kyodai Chiiki Kaihatu, Keikaku to Kiketu* [Gigantic regional development: its plan and results]. Tokyo University Press, Tokyo
7. Japan Industrial Location Center (1969) Report on the large scale industrial development of Mutsu-Kogawara. Tokyo
8. Yoshida F (1999) High-tech pollution in Japan. Environ Econ Policy Stud 2:91–95
9. Sato M (1990) *Resort Retto* [Resort island]. Iwanami Shoten, Tokyo
10. Sasaki M (1990) Regional development and subsidy. In: Miyamoto K (ed.) *Hojokin no Seiji Keizai Gaku* [The political economy of financial subsidy]. Asahi Shimbun, Tokyo, pp 172–208
11. Hobo T (1996) *Naihatuteki Hattenron to Nippon no Nousanson* [Theory of endogenous development and Japanese agricultural and mountainous areas]. Iwanami Shoten, Tokyo, pp 222–223
12. Miyamoto K (1989) *Kankyo Keizaigaku* [Environmental economics]. Iwanami Shoten, Tokyo
13. Ibid.

Chapter 3
Economic Analysis of Waste Management in Japan*

1. Introduction

One of the biggest social problems facing present-day Japan is the issue of waste and waste disposal. Needless to say, many of the safety hazards that have resulted from the siting and conduct of waste disposal sites remain a present danger to the environment and to people. Three particularly troublesome sites are representative of these problems: (1) Teshima Island in the Seto Inland Sea, which is notorious for the illegal dumping of industrial waste; (2) Mitake Town in Gifu Prefecture, which is used as a site for the disposal of industrial waste; and (3) Hinode town in Tokyo, where the safety of municipal waste disposal has been called into serious question. Such instances of environmental pollution are the result of a number of related causes, including the choice of landfill disposal sites, disputes over siting, a shortage of suitable sites, the burden on the local government, dioxin (polychlorinated dibenzo-*p*-dioxins (PCDDs)/dibenzofurans (DFs)) pollution arising from the use of domestic waste incinerators, and even the newly introduced Law for the Promotion of Sorted Recycling of Containers and Packaging.

I therefore need to elucidate the nature of waste disposal in Japan by first considering the relevant issues that must be taken into account if the problems are to be overcome. Japan imports at relatively low prices and consumes in a regularly wasteful manner approximately one-third of the raw material and resources that it uses. Waste is classified roughly into two types: municipal (domestic) waste and industrial waste. The domestic consumption of raw materials and resources amounts to approximately 4 tons of waste (including municipal waste and industrial waste) per person annually, whereas industrial waste in 1994 amounted to approximately 400 million tons, which is between 80% and 90% of the total quantity of waste. Nineteen kinds of industrial waste have been classified, and those that are classified as "injurious to human health and the living environment" are located in specially managed sites. Unfortunately, the 19 specified items do not cover every kind of industrial waste: the list ignores, for example, the pollution of surplus soil.

------

*Originally published in *Economic Journal of Hokkaido University* (Sapporo, 1999) 28:1–27. With the permission of the University.

The official figures given for industrial waste in 1994 have been listed under three headings: sludge 45%, animal excrement 18%, and construction waste 15%. In fact, the amount of construction waste is greater than this because the figures do not take into account the weight of dehydrated sludge. Whereas approximately one-fifth of all industrial waste produced is disposed of, the amount of waste recycled has reached a ceiling of approximately 40%.

In recognition of the differences between disposal facilities, landfill sites are divided into three types. The 1995 figures for these three types of landfill sites are: (1) the least controlled landfill sites (1653); (2) controlled landfill sites equipped with a liner (988); and (3) strictly controlled landfill sites encased in concrete (40).

Recently, dust from shredded automobiles has been dumped in one of the least con-trolled sites, while the latest survey of the Environment Agency has revealed that heavy metals, carcinogenic substances, and other hazardous materials have been detected in one-third of all the least controlled landfill sites. Although a 1997 revision of the Waste Disposal and Public Cleansing Law prohibits the disposal of municipal solid waste (MSW) in sites without a liner or a wastewater treatment facility, old land-fill sites that had been established before the revision have not yet been properly regulated. A 1997 investigation carried out by the Ministry of Health and Welfare established that, at that time, 538 landfill sites (30% of the total) were excavation land-fill sites, and that they were not equipped with either a liner or wastewater treatment facilities. At the same time, evidence shows that even in controlled landfill sites, torn vinyl and gum liners can still cause geo-pollution. This has focused attention on the safety of the final disposal sites.

2. The Illegal Dumping of Waste and Environmental Pollution

The illegal dumping of a massive amount of waste at Teshima Island in the Seto Inland Sea raised the issue of accountability: that is, who should be held responsible for the environmental pollution caused by the illegal dumping of waste and who should bear the cost of cleaning it up? An investigation by the Environmental Dispute Coordina-tion Commission discovered that the waste dumped illegally at Teshima Island con-tained hazardous materials, such as lead, polychlorinated biphenyls (PCBs), and dioxins (PCDDs/DFs). The area of waste covers $460\,000\,m^2$ and 87% of this area falls outside the limits established for the permitted dumping of hazardous waste. At the same time, not only lead and PCBs, but dioxin was also found in the groundwater in quantities exceeding legally permitted levels. Although no obvious traces of pollution in the subsoil or in creatures living on the sea bed have been detected, there can be no doubt that hazardous waste matter will have been leaking through the flow of groundwater into the sea.[1-3]

The origins of the illegal dumping at Teshima Island can be traced back to 1975, when the company responsible applied for permission to operate a hazardous waste-management disposal business. In 1977, the company modified its application so that it could operate an intermediate disposal business in order to cultivate earthworms in the sludge for the production of a soil conditioner. In 1978, the Kagawa Prefectural Government granted the company permission to do this. In 1983, the Prefectural Public Safety Commission also gave the company permission to organize a scrap metal business authorized to haul shedder dust, waste oil, and sludge, as well as to run landfill operations and carry out open burning. The company, however, also carried out open burning of unauthorized waste, and, as the volume of waste being hauled continued to increase, the Hyogo Prefectural Police became suspicious that the company was violating the Waste Disposal and Public Cleansing Law of 1990, which had been designed to end the excessive movement of waste, landfill treatment and open burning.

Although the company complied with the orders of the Kagawa Prefectural Government to remove the dumped waste and install prevention equipment to prevent the outflow of litter, a good deal of waste matter was left behind. Five hundred and forty-nine inhabitants of Teshima Island requested the right to question 21 producers (also called here "generators") of waste, while in 1993 the Kagawa Prefectural Government asked the Environmental Dispute Coordination Commission to mediate between the company and the residents who were asking not only for the removal of waste, but also for compensation for the damage caused. In 1996, the Takamatsu District Court gave a decision in favor of the local residents who had brought forward a case to request the removal of waste and to ask for compensation.

The point at issue is the responsibility of the three bodies who can be held accountable: the producers/generators of the waste, the waste disposal company, and the local government, in this case the government of Kagawa Prefecture.

Because the original Waste Disposal and Public Cleansing Law proved inadequate as a means to counter the waste disposal company's attempts to evade the issue by arguing that shredder dust was not industrial waste because it still contained valuable products, such inadequacies in the Law left room for the administration to interpret its terms in an arbitrary manner. Consequently, the formula "since the law is defective, the government can take no action and the business involved cannot be held responsible" has, naturally enough, simply enlarged and extended the problem.

However, in July 1997, the Kagawa Prefectural Government did come to an intermediate agreement, and admitted that they had made a mistake in giving permission to the waste service company, while expressing their regret at what had happened. At the same time, the Kagawa Prefectural Government made a decision in favor of an intermediate disposal of waste. This, however, far from satisfied the local inhabitants who had asked for the complete removal of waste, nor were the inhabitants pleased that there was no mention of the generators' responsibility for what had happened to affected residents or any recognition that the free offer by the government of their land had been a precondition for the establishment of the disposal facility site. With financial help from the National Exchequer, the Kagawa Prefectural Government proposed to spend more than approximately 15 billion yen for the cleanup.

Although the producers of the waste were well aware that the waste disposal enterprise had been disqualified from conducting its business, the generators of the waste had, nevertheless, hired the unauthorized disposal business because of its lower than average charges. Although the wording of the law is, in this instance, clear enough in order "to force the solvent generators to bear the burden", the biggest problem for the future is how much the generating enterprise will actually be able to pay. Although most of the 21 generators expressed their intention of bearing the disposal cost on the pretext that it was "settlement money", this amounts to nothing more than a contribution to the cost of the cleanup.

3. Revision of the Waste Disposal and Public Cleansing Law

The original law, the Waste Disposal and Public Cleansing Law in Japan, was originally passed in 1971. In 1991, this law was revised. Although the following items were submitted by the Ministry of Health and Welfare Living Environment Council as necessary features of the revision, their aims have remained unrealized:

1. The producer/generator is responsible for waste that is difficult to dispose: afterward, such waste is limited to four kinds;
2. The producer/generator must take responsibility if the waste is dumped illegally;
3. Resource recycling businesses should be subsidized; and
4. The terms of the law must accord with the Basel Convention, which set limits for the transboundary transfer and disposal of hazardous waste.

The more difficult it becomes to find and construct a landfill site, the more urgent have become requests by the public that the Waste Disposal and Public Cleansing Law should be revised. In September 1996, the Special Committee on Industrial Waste in the Living Environment—set up by the Ministry of Health and Welfare—published its report, *The Fundamental Direction of Countermeasures to Cope with Industrial Waste*.[4] The basic problem of industrial waste disposal, said the report, was that waste disposal was caught in a "vicious cycle" (Fig. 3.1): although it was desperately urgent to create new final landfill sites, it was increasingly difficult to find them.

The committee argued that within 2–3 years at most, all existing landfill sites throughout Japan would be full. Illegal dumping, of which there are as many as 350 cases a year, aggravates the already severe state of present-day environmental pollution, while local residents are distrustful of and hostile toward the construction of new landfill sites, even when these sites are authorized. During the past 10 years in Japan, 220 local disputes over the siting of disposal facilities have been registered offi-

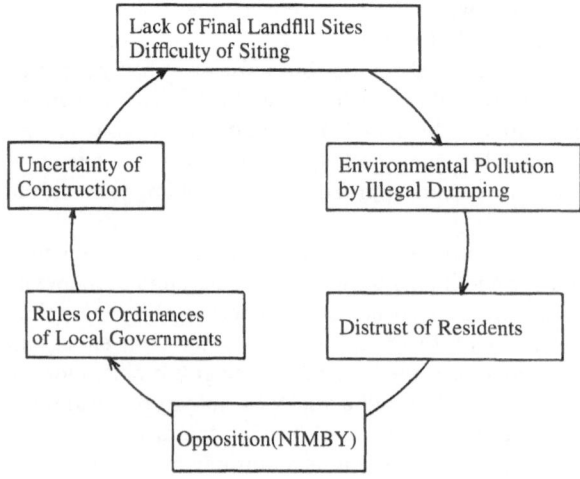

FIG. 3.1. The "Vicious Cycle" of industrial waste disposal Source: Special Committee on Industrial Waste in Living Environment (1996) The fundamental direction of countermeasures to cope with industrial waste. Ministry of Health and Welfare, Tokyo

cially. In October 1997, Professor Masami Taguchi of Rissho University reported the results of his investigations into these cases and came up with the much greater number of 950 disputes.[5] In order to fulfill the ordinances to establish a landfill site, the local inhabitants have to agree but, owing to their suspicion that the construction of the sites will cause a deterioration of the environment, local residents have disputed the proposals (and generally rejected them), thus making it both more urgent and more difficult to construct final landfill sites. The prospect of further construction of sites remains unclear (Fig. 3.1).

The Bar Association of the Kanto District, however, disputed the opinion that the cause of illegal dumping was a shortage of final landfill sites: they insisted that these sites still had considerable capacity for further occupation. In their view, waste was being dumped illegally in order to cut down on the cost of the "treatment fee".[6] The problem, they argued, is money: the waste disposal enterprises seek to increase profits by reducing costs by applying the principle "savings in application of constant capital".

The report of the Ministry Committee suggested the following four measures to counter the "vicious cycle":

1. Promotion of waste reduction and recycling;
2. Improvement of safety standards and a recovery of public trust in the legal treatment of industrial waste: by strengthening the standards, by permitting the creation of mini-landfill sites, and by the full public disclosure of all relevant information;
3. Enforcement of countermeasures against illegal dumping by strengthening the penal regulations and by enriching the existing system; and
4. The establishment of a system to provide the necessary funds to restore polluted land to its original condition, along the lines of the "superfund" for restoration purposes.

The report, however, says nothing whatsoever about the costs to be incurred by the producer/generator of the waste, the disposal business, or the local government.

Meanwhile, the Japan Bar Association produced its own report on the revision of the Waste Disposal and Public Cleansing Law.[7] The report stated that the creation of final landfill sites was not itself sufficient to address the problem: it was necessary to pursue the generator's responsibility as a factor in a society newly committed to the circulation of resources. It would also be necessary to establish a quantitative regulation to an intermediate facility, thus aiming to abolish an lower limit on landfill sites and, so, to establish a closed system for the disposal of hazardous wastes. The report also argued that the generator should be required to donate to a fund for restoration and compensation, and that where illegal dumping had occurred, the local government should have the authority to issue an order to restore sites to their original condition. It also insisted that no waste disposal facility should be sited anywhere near a headwater reservation area, that information about siting should be publicly disclosed, that local residents should have the right to operate a surveillance system to monitor the sites, and that the construction standards for final landfill sites should be reconsidered.

The Waste Disposal and Public Cleansing Law was further revised in 1997. The revision included the following features:

1. Procedures for siting must be clarified through adoption of the EIA process. (The installer of the facility must carry out an investigation into the effect of the facility on its immediate environment, while the local governor must publish the results of the investigation and hold a hearing to consider the opinions of those affected, both government and stakeholders, before permission can be given to construct a landfill site.)

2. The management record of the disposal facility must be made public. (The installer of the disposal facility has to also keep a record of the management and maintenance of the facility and, again, according to certain terms, must make this record public.)

3. The appropriate operation and maintenance of the final landfill site must be secured. (The installer of the controlled final landfill site bears the cost of the operation and maintenance during the period of the landfill.)

4. The penal regulations for the illegal dumping of industrial waste must be strengthened. (A maximum fine of 100 million yen to be imposed on anyone judged guilty of illegal dumping.)

5. Illegally dumped waste must be dealt with appropriately. (The business world and the government must donate/create a fund to cover the expenses for the disposal of an unknown dumper's industrial waste.)

However, the revised law has been criticized on several counts. The responsibility to be borne by the generator of the waste has actually been reduced, rather than extended. The fund to restore land damaged by illegal dumping depends entirely on voluntary contributions. The opinions of the interested groups—the local government, the affected parties, and the specialists—are granted no more than a hearing. No regulations are imposed upon siting. As far as reducing disputes over industrial waste goes, 60% of all local governments consider the revised law useless.[8]

So, why, in spite of all the revisions to the Waste Disposal and Public Cleansing Law, has a radical solution not yet been found? My own opinion is that the existing Law and the general administration of waste control as sponsored by the Ministry of Health and Public Welfare is fundamentally flawed.

Under the direction of the Ministry of Health and Public Welfare, the Japanese Government's administration of the public cleansing services has focused on the prevention of infectious diseases, and, to achieve this, incineration has been regarded as the most effective method. Subsequently, in order to enlarge their market, refuse facility managers took advantage of the public tendency to throw away domestic garbage as nothing more than "obsolete scrap". The resultant cost of disposal, approximately fivefold higher than for the disposal of industrial waste, has been covered by taxpayers rather than by the producers or removers of the waste. (This does not, however, imply that the disposal of industrial waste is perfect either.) As a result of these discrepancies, large-scale incineration operations carried out by the public cleansing service have led to dioxin pollution, landfill sites that are now about to overflow, and very slow progress has been made in recycling efforts.

Japan relies on acquiring raw materials still obtainable from abroad at low prices. While this is so, and while Japan has lax criteria for waste treatment and for the construction, operation, and maintenance of disposal facilities, there has been little need to cut down the environmental load or to conserve depleting resources, and so little progress has been made in the reduction or recycling of waste matter. To overcome this inertia, it has been proposed that, as a measure of ecological financial reform, a tax should be imposed on virgin materials and that the recycling of waste matter should be subsidized. Along the lines of the USA RCRA, superfund (CERCLA), and the German Soil Protection Law (*Bodenschutz Gesetz*), strict criteria for the construction, operation, and maintenance of disposal facilities need to be laid down, while a system to fix the responsibility of the producer/generator must also be established. This has to be done to protect the soil and the groundwater, as well as to reduce and recycle waste.

Once this framework has been put in place, it will be necessary to integrate the present Japanese waste administration systems under a different control from those that operate at present: the Waste Disposal and Public Cleansing Law is under the control of the Ministry of Health and Welfare, whereas the Recycling Law and the Law for the Promotion and Utilization of Recyclable Resources are under the control of the Ministry of International Trade and Industry. One instructive model is a policy run by the OECD, namely Extended Producers' Responsibility (EPR).

Only a business that is able to harmonize the developmental life cycle of a commodity all the way from its design to its recycling—through production, marketing, use, waste collection, transportation, and reuse—will be able to minimize the environmental load synthetically. The second issue will be to decide how the costs are to be borne. Even if, at first, the manufacturer must carry a tentative burden, the ultimate cost, whether it is absorbed by the regular price or shifted to consumers, will depend on the market and the conditions of competition. Problems that will have to be solved independently concern new investment, the barriers erected to prevent new competitors, and the danger of monopolies, all of which will be accompanied by the extended producer's responsibility for bearing the cost of the enterprise.

In Japan, the extended producer's responsibility, seemingly very strict, has been characterized by terms such as "zero emission" to represent zero waste (the attempt by automobile manufacturers to recycle material under the guidance of MITI), and by the urgent enactment of the Law for Recycling Home Electric Appliances.

All these measures are indispensable if the socio-economic system is ever to conserve the environment.

4. Law for the Promotion of Sorted Collection and Recycling of Containers and Packaging

4.1 Point at Issue

Because nearly half the volume of domestic waste in Japan consists of containers and packaging, namely cans, bottles, polyethylene terephthalate (PET) bottles, and paper packages, a law was passed in April 1997 to promote the Sorted Collection and Recycling of Containers and Packaging. However, this new law was immediately contested: the Tokyo Metropolitan Government and 12 other cities called for a reconsideration of the law because of the excessive burden that its provisions would place on local government finances.

As a result of the modern practice of mass production, mass distribution, and mass consumption (and, thus, mass waste), the need to attract customers has led manufacturers to use many kinds of packaging designed to appeal to the buyer, which, in turn, has led to a shocking waste of materials used to make containers and packaging. In addition, as the number of nuclear families increased, so too did the volume of small goods to provide for these families; this has added greatly to the mass waste of containers and packaging.

The point at issue is whether we can seize this opportunity to change the waste disposal system by encouraging recycling and, so, reducing waste by involving the enterprise that, in the first instance, left the waste untreated.

4.2 A Comparison of the Japanese, German, and French Systems

The Law for the Promotion of Sorted Collection and Recycling of Containers and Packaging stipulates a number of conditions: first, a city, town, or village must establish classificatory criteria for the sorting and sacking of different kinds of waste matter according to their composition (i.e., whether the waste is biodegradable or non-biodegradable), the type of sacks to be used by the consumer to collect the material must be specified, and the waste, thus sorted, must be collected at set times. Second, the domestic (and other) consumer must cooperate by sorting and sacking separately the various kinds of waste. Third, the manufacturers and users of containers must ensure that the recycling businesses remarket the material as recycled goods.

Japan enacted this law only after the German and French governments had earlier inaugurated their own systems for the sorted collection of containers and packaging. The new law, sponsored by the Ministry of Health and Welfare, received important support from the All-Japan Prefectural and Municipal Workers Union, a local government labor union, which agreed to cooperate with the policy as a means of maintaining and expanding a directly managed union worker's job.

The German system, Dual System Deutschland (DSD), provides that the collection and classification of waste should not be carried out by the local authority, but by a separate organization that is responsible for recycling goods marked with a green dot, and that the manufacturing enterprises and sellers of goods marked with a green dot should pay the DSD the costs of collection. The French system, on the other hand, requires the local government to take responsibility for the collection of waste, while the responsibility for recycling is in the hands of the Eco-Emballage Company, the operations of which are financed by the container manufacturers and users, who must also guarantee the lowest price and take back recycled goods. Although the Japanese government adopted a system closer to the French system, taking into consideration the present situation in which local governments and inhabitants have a joint responsibility for the recovery of resources, the local governments must also establish the criteria for the sorting and permitted volumes of collected waste, and there is no assurance that the manufacturers will guarantee the lowest price for the goods to be recycled.

4.3 The Heavy Burden on Local Government

The most serious burden imposed on local governments by the new system is the exceedingly heavy cost of collection. For instance, whereas local governments must use public tax to pay ¥25–30 per PET bottle, the manufacturers during the 1997 fiscal year were only expected to pay ¥10 000 per ton, the equivalent of no more than ¥1 per 10 g. The disproportionate rate shows the relatively heavy burden borne by the local government.

While local governments have been earnestly engaged in carrying out the collection of sorted goods, they are also expected to bear an additional heavy financial burden, something in the region of ¥100 000 per ton PET bottles, because two types of industry are exempt from the duties: a manufacturing industry with fewer than 20 employees and an annual income of under ¥240 000 000; and a commercial service industry with fewer than five employees and an annual income of under ¥7 000 000. At the same time, the local government faces supplementary expenses incurred by the sorted refuse criteria: cleaning, condensing, and the prevention of remixing—equal to the expenses of maintaining a 10-ton truck.

For these reasons, Tokyo and 12 other unhappy cities requested that the new law should be reconsidered. Tokyo had, in fact, already enacted its own ordinance (Tokyo Ordinance), which requests the enterprises concerned to make their own voluntary collections over and above those necessarily required by the national Law for Promotion of Sorted Collection and Recycling of Containers and Packaging.

The Tokyo Ordinance has three steps. The first step seeks to promote the enterprise's voluntary collection. The second step, which incorporates the first, requests the distributor to install a collection box and asks all enterprises to collect the container material as a recovery resource. The third step requests the installation of a collection box to be used by the collector and distributor of the sorted material, requests container and content makers to carry out intermediate treatment and recycling, and requests the local government to take temporary responsibility for the delivery of other material to the intermediate disposal facility. The 12 other cities that supported the protest asked for the same voluntary collection on the part of the recycling enterprises. In October 1997, the Tokyo Municipal Government and the makers of PET bottles reached an agreement that the PET makers should take responsibility for establishing a recycling facility and that the Tokyo Municipal Government should rent out property to PET manufacturers for storing collected goods before onward transportation.

4.4 Weak Incentive to Waste Reduction

The German DSD requires the enterprise to pay the cost of container and packaging disposal. In the case of plastics, for instance, this cost is approximately 20-fold higher than the rate for the disposal of glass: this is intended to discourage the manufacture of containers and packaging that are difficult to dispose of. In Japan, on the other hand, thanks to the increase in sales of mineral water, sales of plastic bottles have rapidly increased (by more than threefold between 1996 and 1997); indeed, in expectation of the Law for the Promotion of Sorted Collection and Recycling of Containers and Packaging, the makers of PET bottles actually lifted the voluntary self-control on the manufacture and use of such bottles. It is a great weakness of the Japanese system that nothing is done to prevent or discourage this kind of thing from happening. The cause of this weakness is attributed to the policy priority placed on the recycling process itself, but not resource reduction, without fully considering cost allocation.

The Law for the Promotion of Sorted Collection and Recycling of Containers and Packaging stipulates that returnable glass bottles are to be collected by the enterprise but, in fact, the bottles are not reused immediately; in Sapporo, they are smashed into cullet (gobbets of glass that can then be remolded). This is part of a scheme instituted by the Sapporo Municipal Government that asks residents to sort cans, glass bottles, and PET bottles separately from "inflammable garbage". Residents pack the sorted waste into specified transparent bags, which are collected by regular garbage trucks once or twice a week. At the same time, the majority of the rapidly increasing number of PET bottles on the market becomes pure waste because only approximately 30%, at the highest estimate, is collected: therefore, the absolute volume of PET bottles that are neither recycled nor reused has also increased, just as rapidly.

4.5 The Problem of a Mismatch

Currently, excessive stocks of used paper are burnt because of a shortage of storage space (stockyards). A similar imbalance between supply and demand is likely to occur in the recycling of containers and packaging. The new act provides that competent ministers of the Finance Ministry, MITI, the Ministry of Health and Welfare, and the Ministry of Agriculture, Forestry, and Fisheries should decide on the recycling program, whereas city, town, and village authorities must establish plans for sorted collection. Plans for 1997 through to 2001 have been drawn up, and it appears that no more than 15% of cities, towns, and villages throughout Japan are proposing to collect PET bottles. Even under the recovery program operating in 1997, no more than 44% of PET bottles sold had been recovered.

With regard to plastics (polystyrene foam, vinyl chloride, wrapping foil, and such like)—which, from the 2000 fiscal year, are to be collected separately—there is no likely technology available for their recycling at present. Even Germany faces the problems of coordinating supply and demand, the development of technology (R&D), and the product market–development of recycling, and this will constitute a major task for the future. It is particularly urgent to prepare stockyards (that is, to allocate storage space) to house sorted and collected container and packing waste.

4.6 Further Problems and Prospects

It is difficult to argue that the Law for the Promotion of Sorted Collection and Recycling of Containers and Packaging will offer an opportunity to reform the system for the mass disposal of waste. Indeed, all it entails is "mass recycling at public expense".

So, what shall we need to do in order to promote the use of returnable bottles and put a stop to the wasteful habit of using most containers no more than once? If we could actually do something about this, it would be the key to overcoming the problem of mass waste. Europe has already devised a combined method of "command and control", and has created the economic instrument to implement it effectively. The Environment Agency investigated this method in its report *Committee on the Promotion of the Re-use of Containers and Packaging*[9] (in August 1996, the Chairman of the Committee was Professor Kazuhiro Ueta of Kyoto University). The committee made the following recommendations:

1. That, as in Germany and Denmark, manufacturers should be obliged to use return-able containers and should be forbidden to market containers that are non-returnable;
2. That, as in Finland, the use and sale of any containers that are non-returnable should be taxed;
3. That, as in Germany, the quality of returnable containers should be standardized;
4. That, as in The Netherlands and Germany, the introduction of a system for the return of PET bottles should be promoted; and
5. That a system for the deposit and refund of returnable containers should be introduced.

It is already clear from the European experience that these arrangements have practical as well as theoretical effects. What then needs to be considered are measures that will be required to support these arrangements: the construction of stockyards for reusable goods, better rinsing facilities, and the promotion of this issue as a priority at the level of public institutions.

An investigation by the German Federal Ministry of the Environment has demonstrated that a smaller environmental load is imposed by the use of returnable bottles. The Ministry of the Environment found that a returnable bottle consumes less raw material than a non-returnable bottle, that it emits fewer pollutants, and that waste is reduced if the recycling process is kept within a radius of 100 km and if the product is reused 25 times.[10]

However, the Law for the Promotion of Sorted Collection and Recycling of Containers and Packaging is likely to cause problems because mass recycling will be carried out at public expense. To overcome this weakness, it will be necessary to link the economic measures taken with various methods of direct control. European countries have established such a system and excellent results have already been obtained.

5. Economic Analysis of a Garbage Charging System

5.1 Point at Issue

The notion of a "garbage charging system" has been attracting public attention as a means of reversing the accelerating increase in quantities of household refuse and municipal waste. An investigation carried out by the Ministry of Health and Welfare in June 1993 into the garbage collection arrangements of 3236 municipalities found that 1317 local governments had established systems for the controlled collection of waste, which seemed to correspond with the number (40%) of local governments that had introduced a "payment per unit weight system".

Three types of system have been considered. The first fixes the cost by quantity, which is the strict "payment per unit weight system"; the second charges only for large quantities, which means that, up to a certain weight, collection and disposal are free of charge; and the third levies a fixed charge, regardless of the amount of garbage to be disposed of. When the charge is fixed according to quantity, four other conditions may be imposed: (1) residents must buy a properly designated garbage bag; (2) residents must buy an appropriate sticker label or seal and attach it to the proper garbage sack; (3) the charge may depend on the size of the litter bucket; or (4) each household must pay a lump sum per person per month.

If carried out effectively, these measures would have the following effects: (1) the reduction of waste; (2) the promotion of recycling; (3) the spreading of the cost burden impartially according to quantity; (4) the avoidance of mixing different types of waste; (5) the raising of financial resources; and (6) awakening of the public consciousness. The measures, though, may also create problems: (1) residents may find themselves having to pay a double levy; (2) the measures may appear to residents to be a form of retrogressive taxation; (3) the measures may act as an incentive to illegal dumping; and (4) the measures may not always act in harmony with, or may come into conflict with, the overall reform of mass consumption and mass waste. It may, therefore, be better if the system was first tried out as a charging system for ordinary business waste.

5.2 Household Garbage Disposal Service

Opinions differ about the kind of service that the household garbage disposal system should offer. Should it be a public service, or should it operate as a private enterprise in the open market? Or should it combine elements of both public and market-operated services? A survey of the present household garbage disposal service and of public opinion should be carried out, bearing in mind all the points mentioned by Yamakawa and Ueta[11] and listed above.[12,13]

I should like to think of this issue as a specific problem arising from a historic change in the nature of household garbage, and the effect that this has had on local governments' garbage disposal services. In the days when household garbage consisted mostly of food waste, the local government disposed of it as a matter of public health, in an effort to prevent disease and to discourage illegal dumping. These days, however, household garbage consists of many different kinds of waste, quite apart from food waste. The new Waste Disposal and Public Cleansing Law (within the jurisdiction of the Ministry of Health and Welfare) has been enacted to cover the unprecedented situation that society now faces.[14]

With a change in Japanese people's lifestyle and an increase in mass consumption, there has been an increase in the production and consequent wastage of paper, containers, and packaging. A survey carried out by the Sapporo Municipal Government, for example, concluded that the main reason for an increase in flammable garbage has been the increase in waste paper and plastics. Consequently, what worked a generation ago no longer works now, and changes in the practice of garbage disposal have to be made to correspond with the realities of present-day living. Because the new types of garbage can all be recycled, and may be reused in alternative ways (other than by disposal through traditional methods of incineration and landfilling), the disposal business can now be considered a marketable service, its charges corresponding to the income level of respective households. Consequently, we can say that the nature of the present household garbage disposal service reflects the characteristics of both public and marketable services.[15]

It has been argued that the free public service of garbage disposal has brought about an increase in the amount of garbage to be disposed of. Because disposal costs can, at present, be left to local governments to finance, there has been no need for enterprises to restrain the ever-expanding manufacture and use of non-returnable containers.[16,17]

Although water supply and sewage disposal are matters of public health and have been deemed issues of public service, certain public charges, corresponding to the amount of water consumed or water wasted, have, in certain places, been levied, and we shall need to find a strong reason in principle to argue that the household garbage disposal system should be run as a government enterprise, free of charge. In certain foreign countries, notably the USA, not only is the water supply managed as a marketable service, but so, too, is waste disposal. Services, whether public or private, and the cost of running them depend on the history and character of each individual local government.

5.3 The Income and Substitution Effects of Charging the Public for the Disposal of Household Waste

The imposition of a charge for the collection of household waste would be the equivalent to raising the price of household goods. We can thus consider the consequences of introducing a charge system in terms of the income effect and the substitution effect.[18,19] The income effect would be a reduction in the generation of solid waste as a result of a reduction in spending on goods. For individuals within the high-income bracket, a waste disposal charge would have little effect and there would be no reduction in the amount of garbage they threw out. However, people within a fixed-income bracket, such as pensioners, or those with a low income, would find the charge eating into their savings and they would try to reduce the amount of garbage that they had to dispose of. This situation has been called "the retrogressiveness of the waste charge", and any waste charge aimed at having an income effect would do little to reduce waste: it would, indeed, lead to the problem of retrogressiveness. The substitution effect, on the other hand, aims to reduce waste by finding ways other than waste collection as a means for its disposal. These substitutive ways include using food waste as compost, on-site incineration, reuse, recycling, and illegal dumping. Both on-site incineration and illegal dumping, however, would only add to the burdens already imposed on the environment.

A case study has, in fact, been made of the income and substitution effects caused by the introduction of a waste charge, and we should now carefully consider the results of this study, outlined below.

5.4 The Waste Reduction Effect of a Garbage Charging System

For several years, a group led by Professor Nobutoshi Tanaka of the Faculty of Engineering, Hokkaido University, has been conducting a nation-wide survey of the problems that result from the imposition of a household waste disposal charge.[20-22]

The survey reveals that approximately 40% of all local governments have adopted a garbage charge aimed at reducing the cost of disposal, whereas other local governments have adopted a designated bag system, which aims to reduce waste by having it sorted completely before it is collected. Data provided by 18 local governments make it clear that, in spite of great differences between the amount of waste collected in different local government areas, there had been a tendency for a gradual increase in the amount of garbage before the imposition of the charge, and a tendency for the amount of garbage to decrease after the charge had been levied. After the enforcement of the waste charge system, there was a rapid reduction in the volume of waste produced but, after approximately 3 years, the volume of garbage leveled out, or even started to increase again.

1. In Moriyama City (Shiga Prefecture), Date City (Hokkaido Prefecture), and Oshamambe Town (Hokkaido Prefecture), the amount of garbage dropped by 30%–50% and, thereafter, the percentages remained constant.

2. In Ina City (Nagano Prefecture) and Matsu-ura City (Nagasaki Prefecture), the rate in the reduction of waste has remained more or less constant at approximately 5%.

3. In Hitachi Ohta City (Ibaragi Prefecture), Yuzawa City (Niigata Prefecture), Chino City (Nagano Prefecture), Takayama City (Gifu Prefecture), and Izumo City (Shimane Prefecture), the rates of reduction in the amount of garbage have been as much as 20%, but because the charges were only levied 1 or 2 years ago, the survey must continue to monitor these rates for the time being.

4. In Shiroishi City (Miyagi Prefecture), Dazaifu City (Fukuoka Prefecture), and Tsukushino City (Fukuoka Prefecture), there has been a 5%–10% drop in the garbage rates but, again, we must keep an eye on these rates because data only exist for the 1 or 2 years since the charges were levied.

5. In some cities, however, where the rates of reduction fell initially by as much as 20%, the volume of garbage has gradually increased again, and has now reached the levels recorded before the charge was enforced: places where this has been observed to occur include Taku City (Saga Prefecture), Mobara City (Chiba Prefecture), Naruto City (Yamaguchi Prefecture), and Masuda City (Shimane Prefecture).

The survey has found that a waste charge system makes it possible to reduce household waste by at least 10%, equal to approximately 100 g per person per day. It is significant, however, that, in some localities, rates returned to normal after a small drop in the volume of garbage, which indicates that the charged bag system has no obvious effects on the quantity of non-combustible garbage. However, remarkable reductions in household waste were observed when ordinary business waste was separated from collected household waste. The estimated average reduction of household waste amounted to 87 g per person per day, which worked out at an average 15% in the rate of reduction. The contribution made by efforts at self-disposal came to 50%, the contribution made by a reduction in "obsolete scrap" was 40%, whereas the recovery of resources contributed 10%. The average rate of "refusing to buy what is likely to be wasted" stood at 21%, which indicates that "the charging system" has no influence on citizens' refusal to buy unnecessary goods.

In view of these particular substitution effects, it appears that the charging system leads to a reduction in waste as a result of an increase in garden compost and primary incineration. The self-disposal of kitchen waste as compost is clearly appropriate, but backyard burning is likely to emit a dangerous quantity of chlorinated chemicals.[23]

The survey confirms that in areas where the local government has adopted the charged bag system, inhabitants have made their own evaluations of the system, and have modified their behavior accordingly. In particular, if residents believe that the financial burden has been lightened, they feel ". . . a sense of achievement that they have succeeded in reducing waste", whereas the introduction of the system ". . . has raised the residents' concern about the problems of garbage and its harmful effect upon the environment". At the same time, ". . . the charging system appears to have caused few cases of illegal dumping".[24] Because most families were already partici-pating in programs of resource recovery at the time of the levying of the waste charges, resource recovery has contributed little to new figures of waste reduction.

In a large city where the waste charge has not yet been imposed, an opinion poll revealed that half the inhabitants believe that the quantity of waste reduction will depend on how the charge is levied. Most are of the opinion that the charge should depend on the volume of solid waste to be disposed of, and that charges should not be levied for an average volume of waste, but only on waste that exceeds the average. An average volume of waste should be collected free of charge and a charge should be levied on whatever exceeds that average volume. This would have a bearing on the income effect of the waste charge. It also indicates that people wish to clarify waste disposal expenses, believing that a certain minimum of waste should be collected and disposed of at public expense, and that only excessive waste should have to be paid for by the individuals responsible for that waste. This would eliminate the problem of retrogressiveness.

5.5 Conclusion

We have now seen that the most we can expect a waste charge to achieve would be a reduction of waste in the region of 10%. Anything more than this would be exceptional. As Dr. Tanaka emphasizes, however, the waste charge is important as a means of raising people's consciousness of the environment and increasing their awareness of the dangers that the environment faces. It is therefore important that the public should be informed about the environment and notified of the hazards to which it is subjected. At the same time, reasons for the charges levied on the collection of waste must be care-fully explained, while details of accounting must be made publicly available.[25,26]

When we consider the income effect of the waste charge, we must also consider the related question of retrogressiveness. Izumo City and Takayama City appear to have solved the problem by using a "penalty system" and charging only for large quanti-ties of waste: a certain number of bags and seals are distributed free of charge and residents have only to pay a collection charge for any waste over and above the stan-dard volume. This scheme has the support of most of the population.

At present, the aim of the substitution effect is to prevent waste emission by encour-aging self-disposal and recycling. Self-disposal entails, in the main, turning food waste into compost and burning one's own domestic refuse. Some local governments sub-sidize these endeavors. In big cities, however, where few people have gardens, making compost is a problem, while on-site burning is likely to be a source of air pollution. Although systems for the recycling of recovered resources have already been estab-lished, they are quite separate from systems that charge for waste collection, and they differ widely from city to city. In addition, here we can say that the greater the number of systems used for recovering resources and sorting materials that a city adopts, the greater the rate in the reduction of waste that such a city is likely to achieve.

When Dr. Yasoi Yasuda carried out an investigation in Noda City (Chiba Prefecture), he found that resources were sorted into 14 different categories and that, as a consequence, the amount of non-combustible domestic waste fell by 38%, whereas Moriyama City (Shiga Prefecture), which has, for a long time, directly managed a resource recovery program, has achieved a reduction rate of 59%. These figures confirm the proposition, suggested earlier, that waste reduction can be best achieved not by a waste charge, but by a complete and thorough program of rigorous sorting and separate collections. In order that the public should fully participate in the sorting and collecting of resources, it is essential that the local authority, the residents, and the disposal enterprises should work together as a united body, in which everyone is properly and fully informed about the environment and the local infrastructure (such as the location of stockyards) through the distribution of suitable explanatory publications to each household.

6. Pollution by Chlorinated Dioxins and Related Compounds

Japan is now more polluted by dioxins and related compounds than any other country in the world. The rate of dioxin generation by the waste incineration facilities of each local government authority is now as much as 5 kg per year, and an investigation is currently being conducted into the actual conditions of the most obvious sources of this generation: those incineration facilities that dispose of chlorinated organic compounds, waste oil, and medical waste from hospitals, as well as iron manufacturing plants. In the neighborhood of these facilities, the concentration of dioxin in the air has been recorded as 0.6 pg toxicity equivalents (TEQ)/m³ in big metropolitan and residential areas, compared with background areas without artificial pollution with average concentrations of dioxin in the air of 0.05–0.06 pg TEQ/m³ (see Table 3.1). "Toxicity equivalents" indicate the sum of toxic values after the toxicity of each dioxin homologue has been converted to 2,3,7,8-tetrachlorodibenzo-p-dioxin (TCDD), which is the most toxic of all dioxins.

Polychlorinated dibenzo-p-dioxins, known simply as PCDDs, are, in general, often dealt with as kinds of dioxin, and we may add polychlorinated dibenzofurans, or PCDFs, to this group. The toxic mechanism of coplanar PCBs is similar to that of dioxins. Dioxin has been notorious ever since the Americans used it as a defoliant during the Vietnam War; in Japan, it is well known that PCDFs and coplanar PCBs are responsible for Kanemi Oil disease.

It is said that 70% of all the world's garbage incinerators are located in Japan and that the incinerators that burn MSW are the principal originators of dioxin. Although no complete investigation has yet been made of the incinerators and small furnaces that burn industrial waste, there can be no doubt that the pollution for which they are responsible is likely to be intensely hazardous. Nor can there be any doubt that the present high level of dioxin pollution in Japan's environment is an inevitable result of the Japanese principle of waste incineration.

Although Professor Ryo Tatsukawa of Ehime University detected dioxin in a sample of fly ash taken nation-wide from nine incinerators as long ago as 1983, it was not until 1997 that the Ministry of Health and Welfare finally decided to make public the

TABLE 3.1. Comparison of polychlorinated dibenzo-p-dioxins/dibenzofurans in the air between Japan, Europe, and the United States of America

Japan (Air Pollution Control Division 1996)

Area		Average concentration	(pg TEQ/m³)
Residential area near industrial district	1990	0.10–2.3	0.57
	1992	0.01–2.0	0.62
	1994	0.01–2.6	0.63
Metropolitan areas	1990	0.00–4.7	0.66
	1992	0.00–2.6	0.60
	1994	0.00–3.0	0.37
Middle-sized and small cities	1990	0.00–4.1	0.71
	1992	0.00–1.9	0.36
	1994	0.00–1.1	0.20
Background areas	1990	0.00–1.2	0.19
	1992	0.00–0.03	0.01
	1994	0.00–0.11	0.02

Europe and the USA

Area	Country and area	Years of measurement	Concentration (pg TEQ/m³)
Rural area	US, Minnesota, rural (Maisel & Hunt 1990)	–1990	0.021 ($n = 21$)
	Germany, Kiberg, rural (Liebel et al. 1993)	1990–1992	0.04
	Germany, rural (Fiedler 1994)	–1993	0.025–0.070
	Sweden, country (Broman et al. 1991)	1989	0.004
	The Netherlands, background (Bolt & de Jong 1993)	1991	0.01–0.015
Urban area	US, Connecticut, urban (Maisel & Hunt 1990)	–1990	0.092 ($n = 27$)
	US, California, urban (Maisel & Hunt 1990)	–1990	0.091 ($n = 34$)
	Germany, industrial area (Liebel et al. 1993)	1990–1992	0.15
	Germany, urban (Fiedler 1994)	–1993	0.070–1.350
	Sweden, suburb (Broman et al. 1991)	1989	0.013
	Sweden, city (Broman et al. 1991)	1989	0.024
	UK, city (Clayton et al. 1993)	1991	0.04–0.10
Near sources	Germany, close to source (Fiedler 1994)	–1993	0.35–1.60
	The Netherlands, MSWI (Bolt & de Jong 1993)	1991	0.01–0.15

Source: Sakai S (1998) Environmental policy and status for chlorinated dioxins and related compounds. Environ Econ Policy Stud 1:166

actual situation of dioxin emissions from Japanese waste incineration facilities. Acting on Professor Tatsukawa's findings, the Ministry did set up a special investigating committee, and, although in 1990 it established guidelines to control PCDDs/DFs, these guidelines had no legal status. Consequently, both pollution of the air by dioxin emitted from incinerators and the dumping of fly ash in controlled landfill sites were left to take their own course.

When we reflect that Japan incinerates more waste matter than any other country in the world, we really should have paid much more attention to the results of reports on levels of dioxin in the environment. Yet, in spite of our being the country with the world's highest levels of air-borne dioxin, the measures Japan has taken to control pollution caused by emissions from waste incinerators lag far behind those of other countries. In June 1996, the Ministry of Health and Welfare finally set the limit for the daily intake of dioxin that would be considered harmless to human health at 10 pg TEQ/kg/day. At the same time, the Ministry's revised Air Pollution Control Law Enforcement Ordinance and the Waste Disposal Law Enforcement Ordinance set out to regulate standards for the emission of dioxin from incinerators: in the case of a newly built incinerator, less than 0.1–5 ng TEQ/Nm^3 (1 ng is equal to one-one billionth g or ppb); and, in the case of an old incinerator, 1–10 ng. These levels were to be achieved over a period of 5 years, or 80 ng to be achieved over a period of 1 year.

Dioxin can affect the human body in a variety of ways, from acute to chronic toxicity, carcinogenicity, and changes to the thyroid gland. It has recently been noted that dioxin can also induce reproductive toxicity. In the past, dioxin has usually entered the human body via consumption of fish, particularly shellfish, so that it is especially alarming to learn that a baby can absorb dioxin through its mother's milk. In fact, a fat concentration of 51 pg dioxin was found the milk of one mother in Osaka, the highest concentration of dioxin yet noted in a Japanese subject. It has also been suggested that there may be a link between dioxin in a mother's milk and cases of atopic dermatitis: the rate of atopic dermatitis is higher among babies fed breast milk than among babies fed with milk formulas.

A further consequence of dioxin in the environment may be the high death rate of babies in the Saitama District, where many industrial waste incinerators are located, and where the local government has subsidized small-scale incinerators for longer periods than is common in other areas (as reported on Asahi Television's "News Station", March 31, 1998). At the same time, inhabitants of Ryugasaki City and Shintone Town, both in Ibaragi Prefecture, appealed to the court for a shutdown of incinerator operations after it was revealed that the high rate of deaths from cancer could be related to high levels of dioxin in the atmosphere. Such evidence calls for prompt action and a quick solution: a special committee must make a thorough investigation in areas close to waste incinerators in order to trace the connection between states and levels of environmental pollution and injuries to human health.

The most fundamental measure necessary to reduce dioxin pollution must be a reduction in the amount of waste incinerated. During incineration, it will be necessary to control completely, by means of the 3Ts (temperature, time and turbulence), any production of dioxin. Although the mechanism by which dioxin is produced is understood in general, certain problems still await solution: more needs to be known about the parts played by vinyl chloride, waste home electric appliances, waste wood, chlorinated fire retardants and antiseptics, products that include copper (which becomes the catalysis of dioxin in a re-composite reaction), and salty waste foods.[27]

If plastic waste was to be strictly sorted and perfectly collected, if all food waste was to be reduced to compost, and if discarded domestic electrical appliances were to be strictly recycled according to the provisions of the Sorted Collection and Recycling of Containers and Packing Law, then there would be a considerable reduction in the amount of waste needing to be incinerated. In fact, Sweden and Germany have placed a temporary moratorium on the building of new incinerators, and, for newly constructed incinerators, have set a strict standard of 0.1 ng emissions of dioxin: these steps, as well as reducing dioxin, have also succeeded in reducing waste.

The measures that the Japanese Ministry of Health and Welfare have taken to control dioxin emission have actually had the effect of enlarging the market for the makers of waste incinerators, which now centers on the high-technology control of the dumping of waste. Consequently, a considerable financial burden has been placed on the shoulders of the local governments, a burden that, these days, includes expenses for the disposal of fly ash. Most local governments are beginning to demand that measures should be taken against the emission of dioxin from small-scale incinerators, although in an area such as Hokkaido, where the population is dispersed widely and there is a limit as to how much waste can be collected, small-scale local incineration is likely to remain a problem that may be difficult to resolve.

The incineration of waste is responsible not only for the emission of dioxin, but also for the release of toxic gas and carbon dioxide, which runs counter to the aim of reusing resources in the most effective way. If we hope to reduce waste, it will be necessary to put a stop to the old traditional ways of incinerating waste; in addition, we must incinerate what waste remains while ensuring that the emission of PCCDs/DFs is thoroughly controlled. In recognition of the need for social and technical measures to counter the threat of dioxin, the Tokyo Metropolitan Government is about to reconsider its policy "to build an incinerator in every ward", and will give preference instead to the building of recycling facilities.

7. The Transboundary Movement of Hazardous Waste

A further problematic issue in the world of waste is the transfer (as export) of leftover aluminum ash from Japan to North Korea and the Philippines. In recent years, the practice of transferring waste on an international scale has increased greatly, some of it being moved from one corner of the world to another, yet definitions of this transboundary movement of hazardous waste vary from country to country. The OECD has published export and import statistics for the movement of hazardous waste among its member nations (Table 3.2), and its figures for the 5 years from 1989 to 1993 show that Germany exported 2.9 million tons of waste, Belgium exported 1.39 million tons, and The Netherlands exported 0.91 million tons. Belgium was also the largest receiver of waste, importing 3.66 million tons.

Meanwhile, Greenpeace has collected official and unofficial data for the transfer of hazardous waste from member countries of OECD to non-member countries for the years 1989 to the spring of 1994.[28] Germany exported waste on 214 occasions, the USA exported waste on 208 occasions, and the UK exported waste on 101 occasions. In Eastern Europe, Russia imported waste on 300 occasions, the Asia-Pacific imported waste on 250 occasions, and Latin America imported waste on 150 occasions. The USA headed the list of countries who exported hazardous waste to the Asia-Pacific region,[29] at more than 5 million tons exported, with Canada second, and the UK third.

TABLE 3.2. Transfrontier movements of hazardous wastes by OECD countries (1989–93)

(Unit: ton)

(a) Export

Fiscal year	1989	1990	1991	1992	1993
Austria	86 773	68 162	82 129	70 023	83 998
Belgium	176 983	491 784	645 636	37 278	34 073
Canada	101 083	137 818	223 079	174 682	229 648
France	n.d.	10 552	21 126	32 309	78 935
Germany	990 933	522 063	396 607	548 355	433 744
Netherlands	188 250	195 377	189 707	172 906	163 180
UK	0	496	857	0	0
USA	118 927	118 416	108 466	145 556	142 709

(b) Import

Fiscal year	1989	1990	1991	1992	1993
Austria	50 981	19 180	111 595	79 107	28 330
Belgium	1 036 260	1 070 496	1 021 798	208 052	236 010
Canada	150 000	143 811	135 161	123 998	173 416
France	n.d.	458 128	636 647	512 150	342 538
Germany	45 312	62 636	141 660	76 375	78 219
Netherlands	88 400	199 015	107 251	250 355	236 673
UK	40 740	34 983	54 074	44 673	66 294
USA	n.d.	n.d.	n.d.	n.d.	n.d.

Source: OECD (1997) Transfrontier movements of hazardous wastes, 1992–93 statistics. OECD, Paris

South Korea imported far and away the largest quantity of hazardous waste, followed (although a long way behind) by India. An overwhelmingly large proportion of the hazardous waste as a whole consisted of lead, copper, and tin; other types of hazardous waste included sludge, used plastics, used lead batteries, used computers, disused leather, and similar goods.

In 1989, the Basel Convention adopted a proposal to prohibit the transboundary movement of hazardous waste, and this came into formal effect in 1992. Japan subsequently passed a domestic law modeled on the Basel Convention, but it does not cover all cases specified by the Convention (Table 3.3). The Japanese Environment Agency has published statistics for the country's export and import of hazardous waste for the fiscal years 1994, 1995, and 1996. In 1997, documents were issued to authorize the transfer of hazardous waste: for export 5787 tons (49 instances) and for import 7973 tons (55 instances). The export partners were Belgium, Germany, the USA, Indonesia, and South Korea, and they all aimed to recover certain kinds of metal, particularly copper and tin. However, statistics do not always reflect the actual conditions. For example, The Ministry of Finance's *Monthly Report on Japanese Trade*[30] refers to "waste lead, code number 7802", which seems to consist mostly of lead batteries, and is not to be found anywhere at all in the statistics provided by the Environment Agency. This means that the domestic law based on the Basel Convention does not regulate waste lead.

TABLE 3.3. Export and import of designated hazardous waste in Japan (Basel Treaty Regulation)

(a) Export

Fiscal year	1995	1996	1997
Notice to opponent country	1791 tons[1]	8948 tons[2]	4120 tons[3]
	/8 Cases	/6 Cases	/4 Cases
Export license	474 tons	3730 tons	6398 tons
	/3 Cases	/5 Cases	/4 Cases
Issue of export transfrontier papers	2814 tons	1721 tons	5787 tons
	26 Cases	/52 Cases	/49 Cases

(b) Import

Fiscal year	1995	1996	1997
Notice from opponent country	4179 tons[4]	10660 tons[5]	12466 tons[6]
	/21 Cases	/14 Cases	/17 Cases
Import license	2889 tons	10033 tons	9559 tons
	/19 Cases	/13 Cases	15 Cases
Issue of import transfrontier papers	1163 tons	8722 tons	7973 tons
	/47 Cases	/53 Cases	/55 Cases

1. Importing countries are France, Germany, UK, USA, Korea, and Malaysia, all of which have the purpose of collecting metals such as copper and tin.
2. Importing countries are Belgium, Germany, USA, Indonesia, and Korea, all of which have the purpose of collecting and recycling metals such as copper and tin.
3. Importing countries are Germany, Belgium, USA, Korea, and Indonesia, all of which have the purpose of collecting and recycling metals such as copper, lead, tin, cobalt, tungsten, nickel, and cadmium.
4. Exporting countries are Austria, Netherlands, USA, Philippines, Hong Kong, Korea, and Malaysia, all of which have the purpose of collecting and recycling metals such as copper and silver. However, one of the cases has to be landfill disposal of plating wastewater treatment sludge including lead.
5. Exporting countries are Austria, Australia, Canada, Netherlands, USA, Philippines, Hong Kong, and Malaysia, all of which have the purpose of collecting and recycling metals such as lead and tin.
6. Exporting countries are Austria, Australia, Canada, Netherlands, USA, Korea, China, Hong Kong, Singapore, Philippines, and Malaysia, all of which have the purpose of collecting and recycling copper, lead, arsenic, selenium, and tellurium.

Source: Marine Environment and Waste Countermeasure Section of the Environment Agency (1998)

In 1990, Japan exported the largest volume of waste lead to Taiwan (see Table 3.4 for a breakdown of export of waste lead by country and area) but later, because of a pollution problem, the volume dropped, and, in the 1990s, the volume of waste lead exported to Indonesia increased. In 1994, in an effort to conform to the terms of the Basel Convention, the general volume of waste exported by Japan was reduced, but lately the export of waste has begun to increase again, in particular and in large quantities to South Korea and India. "Waste copper, code number 7404", the export of which amounts annually to approximately 50000 tons, has not been regulated, while over the year 1997 the export of waste plastics amounted to nearly 120000 tons.

TABLE 3.4. Japanese export of lead waste and scrap to countries

(Unit: tons)

Year	1990	1991	1992	1993	1994	1995	1996	1997
Korea	2464	1033	–	34	150	804	5721	9344
China	933	3204	4918	539	15	51	18	20
Taiwan	9586	1748	16	–	–	62	148	–
Hong Kong	2376	1602	245	205	–	–	6	–
Thailand	720	4330	4104	2500	–	–	–	–
Philippines	10	8	33	833	120	–	40	–
Indonesia	8832	13319	11222	3379	–	–	649	956
India	547	39	672	672	–	–	1093	4953
Subtotal	25528	25261	20614	8162	304	1195	8018	15615

Note: A few exporting opponents in each year are not counted in the total.
Source: Ministry of Finance (1998) Monthly report on Japanese trade

The Basel Convention was revised in 1995 to regulate the export of hazardous waste (including material to be recycled), and it established a list of waste (effective from 1997) that member countries of the OECD would be prohibited from transferring to non-member countries. The revision attaches a supplement that states that "... the export of waste which the treaty regards as harmless has been excluded". Consequently, no proper list has yet been confirmed of types of waste that still need to be regulated.[31]

References

1. Hanashima M, Takatsuki T, Nakasugi O (1996) A case study of environmental contamination caused by illegal dumping of hazardous waste. *Haikibutu Gakkaishi* [Waste Manage Res] 7:208–219
2. As to the Teshima Problem, see: NHK (1997) *Yasen no Shikan Nakaboh Kohei* [Commander of the field battle]. NHK, Tokyo, Chapter 4
3. In a series of articles printed during November and December of 1997, the Chunichi Shimbun reported that shredder dust had been illegally dumped in the eel-farm waters at Tawara town, Aichi Prefecture. In October 1997, the same newspaper ran a waste problem campaign under the title "What shall we do about the Garbage Archipelago?" The paper claimed that evidence now exists to confirm that the central part of Honshu has been thoroughly polluted by the dumping of construction waste, the open burning of pinball machines and the misuse of self-disposed landfill sites
4. Special Committee on Industrial Waste in the Living Environment (1996) The fundamental direction of countermeasures to cope with industrial waste. Ministry of Health and Welfare, Tokyo
5. Taguchi M (1998) Asahi Shimbun. January 25, 1998
6. Bar Association of Kanto District (1996) *Haikibutsu Shoriho Kaisei ni Mukete* [Report: toward the revision of waste disposal and public cleansing law]. Tokyo; an investigation carried out by the National Police Agency's Bureau for Safer Living (1997) claims that 65% of criminal instances of illegal dumping of industrial waste are motivated by the need to reduce costs
7. Japan Bar Association (1996) *Haikibutsu no shori oyobi seiso nikansuru horitsu no kaisei ni taisuru ikensho* [Report on the revision of the Waste Disposal and Public Cleansing Law]. Tokyo.

8. The source for this statement is The Nihon Keizai Shimbun, on April 9, 1997. On December 26, 1997, the Director of Water Works and the Environment, on behalf of the Ministry of Health and Welfare ordered local governments to re-examine how far the local by-laws had been applied in practice, over and above those regulations whose first requisite for the granting of permission to contractors for the construction of landfill sites is to seek for and obtain permission from residents in the area before construction work can begin. Approximately 70% of prefectures had made the requisite arrangements for seeking and obtaining local residents' permission to construct sites

9. Environment Agency (1996) Report of committee on the promotion of the re-use of containers and packaging. Tokyo

10. Ibid.

11. Yamakawa H, Ueta K (1996) Concerning the garbage charging system: the attainable level and agenda. *Kankyo Kagakukaishi* [Environ Sci Assoc J] 9:277–292

12. Ms. Y Ochiai suggests six ways by which the expenses of garbage disposal may be borne, if the disposal is charged for in accordance with a per unit weight system:

    In proportion to the emission overall;
    In proportion to the emission volume at each stage of the emission process;
    Emission to be free of charge up to a certain volume;
    A combination of costings with (governmental) support;
    A combination of costings with other kinds of support (the refraction type); and
    A combination of costings according to a Fixed Sum System, the costs charged by quantity

13. See: Maruo N, Nishigatani N, Ochiai Y (1997) Ecocycle society. Yuhikaku, Tokyo, Chapter 3

14. Gunjima T (1995) Thinking about user-charge system and its effects. *Haikibutsu Gakkaishi* [Waste Manage Res] 6:163

15. Kitabatake Y, Nakasugi O (1982) Actual conditions of collecting fees of general waste disposal and analysis of its effects. *Chiiki Kenkyu* [Regional Study] 12:57; the authors suggest that "If at the time of decision-making, the local government considers the external effect of the amount of service supplied, then, as the ratio of private burden of the total costs decreases, so the amount of service supplied increases"

16. Ueta K (1996) Economic analysis of charging system of garbage. Keizai Semin March 1996:218

17. It was included in: Ueta K et al. (eds) (1997) *Kankyo Seisaku No Keizaigaku* [Economics of environmental policy]. Nippon Hyoronsha, Tokyo

18. Gunjima T (1995) Thinking about user-charge system and its effect. *Haikibutsu Gakkaishi* [Waste Manage Res] 6:164

19. Ueta K et al. (eds) (1997) op. cit., pp 219–222

20. Tanaka N (1993) Solution and management plan theory of the generation and circulation structure of solid waste in the urban area. In: Tanaka N et al. (eds) Research reports of 1990, 1991, and 1992 science research subsidy. Sapporo

21. Tanaka N (1995) Research of the effects of waste charging system on the general domestic resource consumption reduction life-style. In: Tanaka N et al. (eds) The research report of 1995 science research subsidy. Sapporo

22. Tanaka N (1996) Research of the effects of waste charging system on the general domestic resource consumption reduction life-style. In: Tanaka N et al. (eds) The research report of 1995 science research subsidy. Sapporo

23. Kamishita T et al. (1996) Environment load of on-site burning in the local government garbage which has introduced the charging system. In: Proceedings of Seventh waste management society. Tokyo, pp 81–83

24. Tanaka N (1996) op cit., pp 4–5

25. Ohno M et al. (1995) Survey of opinions about the introduction of charging system of domestic waste. Urban Cleansing 48:41–42; the report makes three points: (1) that we

need to clarify the purpose of the charging system, or whether it should be regarded in terms of the reduction of the public financial burden, or whether it should concentrate on a reduction in the volume of waste produced; (2) that we must not expect too much reduction in the volume of waste from the charging system itself; and (3) that it is, in fact, difficult to say whether a waste charging system would actually be an efficient social method for reducing the quantity of waste

26. Taguchi M (1998) *Gomi Mondai Hyakka Jiten* [Encyclopedia of garbage problem II]. Shin-Nippon, Tokyo, pp 170–173

27. Miyata H (1998) *Yoku Wakaru Dioxin Osen* [Well understanding dioxin pollution]. Godo Shuppan, Tokyo, p 149; Miyata says that various experiments have made it clear that the contribution of salt to the generation of dioxin in the emission gas is so small as to be, in fact, negligible

28. Greenpeace (1994) Database of known hazardous waste exports from OECD to non-OECD countries. Prepared for the second conference of Parties to Basel Convention, March 21–25. Geneva

29. Greenpeace (1994) The waste invasion of Asia. Greenpeace Australia, Sydney

30. Ministry of Finance (1998) Monthly report on Japanese trade. Tokyo

31. Three kinds of waste are listed:

    List A: Hazardous waste included in the treaty under embargo;
    List B: Waste not regulated by treaty; and
    List C: Material not confirmed waste.
    List B includes unscattered copper debris, mine waste, and waste plastics (standard conformity substances)

Chapter 4
High-Tech Pollution—A Historical Survey of Problems Generated by the High-Tech Industries*

1. A New Type of Pollution—The Fairchild Case

Rick Puppo is a gardener who lives in Los Paseos, San Jose, California. When I visited him one evening, he was watching a baseball game on the television (TV) with his sons. His family is typically American. On the table, however, was a small bottle of heart disease tablets, which his younger son, Brian, has to take everyday.

The story goes back to the end of 1980, when it was discovered that newly born Brian had a hole in his heart. Brian had to have an operation one and a half months after birth. This was followed by three more operations, and Brian was hospitalized for 6 months. Only when Rick Puppo read a local newspaper article did he realize that many children in Los Paseos had been born with heart defects.

The Fairchild underground storage tanks of used organic solvent had leaked and contaminated well #13, which supplied the neighborhood's water. According to the victims' map prepared by the citizens and their attorney, the victims were scattered widely throughout Los Paseos. Red flags on the map represented heart diseases, blue flags showed miscarriages, yellow flags signified cancer, and black flags meant deaths (Fig. 4.1).

In April 1982, the victims brought a suit against Fairchild, IBM, and the Great Oaks Water Company. The number of citizens affected rose to approximately 530.

The problem outlined in the lawsuit focuses on the link between contamination from the Fairchild well and the harm suffered by the victims. Even according to the calculations of Fairchild itself, from April 1977 to December 1981, approximately 58 000 gallons of 1,1,1-trichloroethane (an organic solvent) leached into the well. Stirred into action by the demands of the citizens, the California Department of Health Services carried out an epidemiological study. The study, named *Pregnancy Outcomes in Santa Clara County 1980–1982*,[1] consisted of two parts. The first part compared rates of cardiac defects in the area served by the Great Oaks Water Company

---

* Originally published in *Economic Journal of Hokkaido University* (Sapporo, 1994) 23:73–138. This is a revised version of *High-Tech Pollution* (Iwanami Shoten, Tokyo, 1989) [in Japanese]; although the material may seem a little out-of-date, it is essential background to the present-day issues of industrial pollution [Chapter 5].

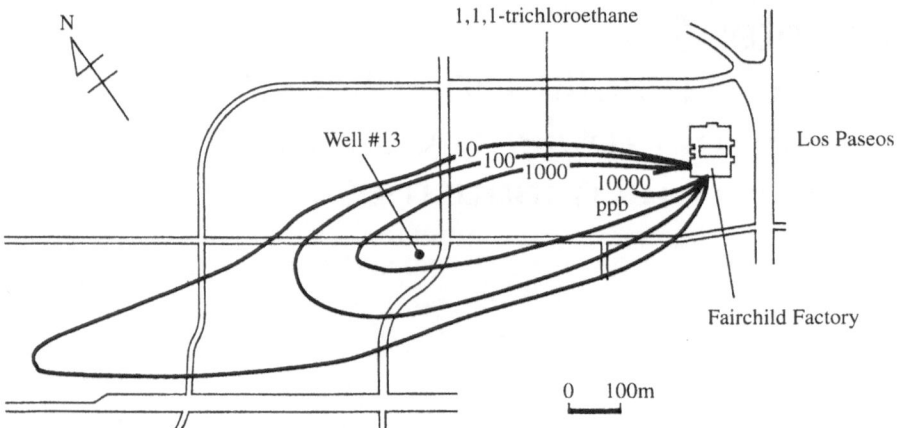

FIG. 4.1. Fairchild Camera and Instrument, San Jose site

TABLE 4.1. Pregnancy outcomes in the Los Paseos area, 1980–1981 (California Department of Health Services, January 1985)

	Miscarriage	Malformation*	Deformation**
Number	41	10	3
Rate	21.5%	6.9%	2.1%
Control area	11.0%	2.2%	0.6%

*Nasolacrimal stenosis, ventricular septal defect, low-set ears, benign form of webbed toes, omphalocele equinovarus, total anomalous pulmonary venous return, imperforate anus and kidney anomaly, Down syndrome, congenital diaphragmatic hernia, and cleft lip and palate
** Left club foot, internal tibial torsion, and right talipes equinovarus

with rates in the rest of the county. The second part of the study investigated the number of cases of spontaneous abortion and congenital anomalies. According to the first part of the study, in 1981, 10 babies with major cardiac defects were born to residents of this district (Table 4.1). This represents an excess of six cases over the expected number. However, the solvent leak is, in itself, an unlikely explanation for this excess.

The second part of the study gave details of the rates of various birth defects: the rate of spontaneous abortions among pregnant women (Los Paseos 21.5%; control area 11.0%); the rate of congenital malformations (Los Paseos 6.9%; control area 2.2%): the rate of congenital deformations (Los Paseos 2.1%; control area 0.6%). However, the investigators concluded that the indirect evidence of the extent and timing of exposure to contaminated water within the Los Paseos area was insufficient to determine whether the leak of chemicals into well #13 had caused the excessive rates in birth defects.

The California Department of Health obviously avoided making a definite state-ment. However, according to the testimony of Dr. S Swan, the person in charge of the study, at the Congressional hearing, the rates of alcohol consumption and smoking, and occupation were not responsible for the differences in the rates of the diseases noted between Los Paseos and the control area. A new report was subsequently released in March 1986. In this report, it announced that, from January to June 1983:

1. Sixty-nine babies were born to mothers living in Los Paseos: none suffered from birth defects; and
2. In the entire Great Oaks service area, 15 of 605 babies had congenital defects. This figure gave the area a birth defect rate of 24.8 per thousand live births, the same number as reported for the whole of the Santa Clara Valley and the San Francisco Bay Area put together.

At the same time, a group at the New Jersey Medical School found a relationship between exposure to 1,1,1-trichloroethane and the appearance of cardiovascular abnormalities in rats.[2]

Against this background, in July 1986, Fairchild, IBM, and the Great Oaks Water Company agreed to pay a "multimillion dollar" settlement to 530 residents. Attorneys and others on both sides of the suit refused to reveal the specific terms of the settle-ment (San Jose Mercury News, July 4, 1986).

This appeared to mean that the Fairchild case was, henceforth, closed. However, in May 1988, a further report, named Pregnancy Outcomes in Santa Clara County 1980–1985,[3] produced by the California Department of Health Services, ruled out 1,1,1-trichloroethane as a cause of miscarriages and birth defects in Los Paseos because residents of an adjoining neighborhood were not similarly affected, even though they had received more of the contaminated water than had Los Paseos residents.

The victims and their attorneys are very critical of this later report. They have pointed out that there has been no health damage since the closure of well #13. They claim that the link between the contaminated water and the birth defects is clear and that the case has already been settled. If I reexamine the final report, for example, I find that the malformation rate (1980, 1981) for the adjoining neighborhood and Los Paseos is 6.0%, whereas that for the unaffected area is 2.2%. In 1982 and 1985, however, the rate of 6.0% decreased to 2.9%. The report contains no analysis of this phenom-enon. With regard to the way in which one may be exposed to the solvent, we have to pay attention to a new theory that stresses the part played by taking a shower or washing one's hands in addition to drinking of the water itself. In Tucson, Arizona, a new study carried out by the University of Arizona Health Science Center found 2.5-fold more cases of heart defects in babies born to people exposed to solvent (trichloroethylene)-contaminated wells near an Air Force base and Hughes Aircraft Company compared with babies born to unexposed individuals.

In Woburn, Massachusetts, a group from the Harvard School of Public Health found a positive statistical association between access to water contaminated with chlori-nated organics and the incidence rate of childhood leukemia. This group also reported that damage to the immune system was manifest because there were altered ratios of T lymphocyte subpopulations.[4]

To sum up, the Fairchild case highlights the problems of a "new type of pollution", because: (1) high-technology industries that seem to be clean are not always clean;

(2) leakage from an underground solvent tank caused pollution of groundwater; and (3) organic solvents, such as 1,1,1-trichloroethane and trichloroethylene, which create occupational safety problems within a factory, are the main pollutants.[5-8]

2. A Warning from Silicon Valley

2.1 An "Advanced Area" of High-Tech Pollution

2.1.1 Silicon Valley

Today, "Silicon Valley" has become a metonymical synonym for any high-technology industrial region. Silicon Valley is the name first given to Santa Clara Valley, Santa Clara County, lying at the southern tip of San Francisco Bay.

Santa Clara County includes Palo Alto at its northern limit and San Jose at its southern end. Approximately 1.47 million people live in this area, while roughly 2900 high-technology related companies employ approximately 230 000 personnel.

After World War II, many computer and electronics companies, such as Shockley Transistor, Fairchild Semiconductor, and spin-off companies (Intel, AMD etc.), moved into this area. Without Silicon Valley, we cannot imagine the post-war development of the American semiconductor and electronics industries. In Santa Clara County alone, these two main industries are responsible for 25% of all USA production of semiconductor and related devices, and for guided missiles and space vehicles. Silicon Valley has approximately 178 semiconductor factories (including wafer manufacturing in 1990), of which 40 factories (22%) employ 250 or more people, 52 factories (29%) employ between 50 and 249 people, and 86 factories (48%) employ fewer than 50 people.

This high percentage of small-scale semiconductor companies is the chief characteristic of Silicon Valley. The rapid growth of the population in this area has led to a shortage of housing and inflation of housing prices, traffic congestion, and a high rate of divorce. Along with these problems, high-tech pollution has added to the difficulties.

As for the labor force, the percentage of professionals is relatively high—30%. On the other hand, the female operatives in printed circuit board manufacturing, for example, are drawn from minority ethnic groups. Hispanics live in the eastern part of San Jose, and Vietnamese live near central San Jose. This is the unknown side of Silicon Valley.

2.1.2 What is High-Tech Pollution?

When I visit Silicon Valley in the daytime, I cannot understand why it is called a "valley". Only when the smog thins out toward evening, and the surrounding mountains appear, can I understand that this area really is a valley. Such pollution, however, is only the tip of the iceberg of the environmental deterioration of Silicon Valley. High-technology attracts enthusiastic attention from all quarters as a leading force of the new industrial society, yet, as the example of Silicon Valley shows, people have gradually come to realize that high-technology, in contrast with its clean image, imposes a heavy burden on the environment. What are the features of high-technology (here,

I focus mainly on the electronics industry and its associates) that relate to these new environmental problems?

1. First of all, "clean" means only "without dust", and the objective is to keep the materials and parts used for making semiconductors "clean". However, the "cleanness" of its actual operation and the environment that surrounds semiconductor factories is another thing.
2. Apart from the apparent waste of water, the waste of gases, the industrial waste, and garbage, which, until now, have been the focus of environmental concern, leakages from "storage" tanks of used solvents is another major cause of pollution.
3. Therefore, besides rivers and the air, groundwater plays an important role as a route for pollution to enter a community.
4. The semiconductor industry uses many sorts of toxic chemicals, gases, and radioactive rays, which have, even in very small quantities, the potential of causing complex chemical pollution.
5. Because high-technology changes very quickly and secrecy prevails in this field, corresponding environmental protection tends not to be forestalled.

Silicon Valley offers a good example of how "high-technology" not only tends to concentrate in one area, but also how it pollutes that area. That is because:

1. Silicon Valley itself is a highly concentrated area of high-technology industries with approximately 2900 high-technology companies, and approximately 230 000 employees, all within a radius of 12 miles;
2. There is the danger of "double exposure": (1) health injury caused by high-tech-related pollution; and (2) occupational exposure to toxic chemicals;[9]
3. In Silicon Valley, there has been extensive and concentrated water pollution, and much money and time is required to reverse this; and
4. A strong civil anti-pollution movement has grown up along with an earnest effort to tackle high-tech pollution, and an earnest effort to tackle high-tech pollution has been made by the local government and the companies themselves.

As a last resort to vitalize local economies, high-technology industries are now expected to move in from all quarters. To forestall a possible "new type of pollution", I would like, in this section, to analyze comprehensively the high-tech pollution of Silicon Valley based on my field research, and to urge that we should listen most carefully to the warnings from Silicon Valley.

2.2 Serious Groundwater Pollution

2.2.1 The Extent of High-Tech Pollution

As I have stated in section 1, the Fairchild case spurred a complete investigation of leakages from groundwater tanks and of the extent of groundwater pollution in Silicon Valley. This investigation found that the pollution is alarmingly extensive. From the data gathered by the "Silicon Valley Toxics Coalition", I would like to offer a picture of the state of pollution in Silicon Valley.

1. Eighty percent of high-risk underground solvent tanks have leaked.
2. Approximately 100 toxic chemicals have been detected in the groundwater.
3. In Santa Clara County, more than 150 leaks from underground tanks have been detected, and over 200 public and private wells have been contaminated.
4. Santa Clara County has 23 USA Environmental Protection Agency (EPA) Superfund sites (extremely contaminated sites where special federal or state funds are used for the cleanup), more than in any other county.
5. Santa Clara County produces 100 000 tons of toxic waste per year and 8.5 million pounds of toxic waste are discharged into the environment.
6. The discharge of 30 000 to 120 000 pounds per year of toxic metals (including silver, copper, cadmium and nickel) into the San Francisco South Bay from Silicon Valley is causing serious environmental damage to the wetlands.[10]

2.2.2 Groundwater Pollution in Silicon Valley[11]

Because people in Silicon Valley take approximately half their drinking water from underground sources, the pollution of these water sources is extremely serious. In October 1981, leakage of the organic solvent 1,1,1-trichloroethane from an underground tank was first detected at IBM. IBM used the solvent for cleaning magnetic disk drive parts, and reported the leakage to the local government authority, which, however, did not take the leakage seriously. If the local government authority had investigated thoroughly at the time, it would have been possible to find the source of the pollution and, therefore, to have stopped its expansion. The pollution had already extended into a 500-ft deep well. Then, in December 1981, the Fairchild Camera and Instrument Company reported that it had lost a significant amount of 1,1,1-trichloroethane; 1 week later, large amounts of 1,1,1-trichloroethane showed up in the Great Oaks Water Company well #13, 2000 ft away. The well water was contaminated with 5800 ppb 1,1,1-trichloroethane, 29-fold the maximum recommended contaminant level set by the EPA. The President of the Great Oaks Water Company, Betty Roeder, admitted that waiting to tell the public about the pollution of well #13 had been a big mistake. She also admitted that the storm water drain from well #13 into Canoas Creek had seeped back underground and had most likely contaminated a well field of the San Jose Water Company. At the same time, she complained that she had never received any help from the California Department of Health Services.

As the investigation progressed, between 1981 and 1982, 21 more leakages from tanks were found, and, until April 1984, 71 leakage points had been reported. Because there had been so much leakage from underground tanks and resultant well contamination, we have to conclude that these phenomena did not occur by accident. Why are there so many underground tanks and why are they so poorly made? Chemical tanks—like gasoline tanks—are located underground because the local fire and architecture ordinances compel companies to put them there. Owners have been happy to obey the ordinance to preserve the attractiveness of the site.

"Bury it and forget it. A Fiberglass fuel tank won't corrode." This is an advertisement for a fiberglass fuel tank; it stresses "Long life", "Savings", and "Fewer Worries". The problem is that many underground tanks are made of the sort of fiberglass that is not only likely to "corrode", but also to crack under uneven ground pressure.

According to the EPA survey of 1986, of 2500 cases of underground tank release in the USA, approximately 50% of leakages were caused by structural failure, while approximately 30% were the result of corrosion. In Silicon Valley, 80% of underground solvent tanks have leaked. In addition to leaks from pipe-linkage, there have been leaks at the filling operations and leaks from the neutralization tank. It is as if ". . . there is a chimney underground," as somebody said. A director of environmental issues at IBM admits that, by pressure testing, it is possible to detect gross leaks from tanks. Originally, however, no one paid attention to the likelihood of leakage. Companies now have to pay the bill for the "saving" of money and superficial "attractiveness".[12]

Factors that affect contaminant transport and the fate of chemicals are: (1) volatilization; (2) gravity flow; (3) dispersion in groundwater flow; (4) interaction with soil or sediment; and (5) transformation.

1. Volatilization means that chemicals such as organic solvents and gasoline may vaporize. The extent of such volatilization will depend on the characteristics of the chemicals, the temperature, and the manner of chemical release.

2. Gravity flow means that liquid contaminants move downward under the influence of gravity. The contaminants can "sink" or "float" depending on whether or not they are denser than water.

3. Dispersion means that contaminants dissolved in groundwater are transported by the groundwater flow. The organic solvent may condense, for example, 2000-fold.

4. Interaction with soil or sediment means that contaminants move with the groundwater due to their interaction with the aquifer sediments. The degree to which specific chemicals interact with the aquifer solids depends on the characteristics of both the chemicals and the aquifer solids.

5. Transformation means that some contaminants may be transformed by biological or chemical reaction, yielding either harmless or more or less harmful products. In the latter case, for example, 1,1,1-trichloroethane is transformed into 1,1-dichloroethylene (*Ground Water and Drinking Water in the Santa Clara Valley, A White Paper*, by California Department of Health Services 1984, p. A-4).

According to a report produced by the Semiconductor Industry Association (SIA) in 1985, while excluding the dilute waste acid stream, the semiconductor industry generates 55 million pounds of hazardous waste nationwide; 13 million pounds are generated in the State of California alone. At the same time, the semiconductor industry generates 29 billion pounds of waste acid nationwide, 32% of which finds its way into deep wells. Eighty-five percent of 44 000 pounds of gallium arsenide waste produced by the semiconductor industry has been dumped into a landfill. Therefore, the high-technology industry itself has become a generator of hazardous waste.

2.2.3 Hazards Posed by Drum Recycling

The community meeting held at Kelly Park's Leininger Center, San Jose, looked like an international conference because the organizer of the California Department of Health Services was a Japanese-American, the engineer was a Chinese-American, and the citizen participants were black, white, Hispanic-Americans and Japanese-Americans.

TABLE 4.2. Contamination level at the Lorentz Barrel and Drum site

Primary Contaminant	Highest level (ppb)	EPA MCL (ppb)	State action level (ppb)
Trichloroethylene	2108	5	5
Vinyl chloride	1100	1	2
1,2-Dichloroethane	270	5	None
1,1-Dichloroethylene	160	7	6
1,1,1-Trichloroethane	220	200	200
PCB	6.4	None	None

Highest level detected is the highest concentration of contaminant found in a groundwater sampling on and near the site

EPA MCL, proposed maximum contaminant level set by Environmental Protection Agency

State action level, drinking water action levels recommended by the California Department of Health Services set unenforceable water quality standards

PCB, polychlorinated biphenyl

The community meeting held on September 3, 1987, was called to discuss the pollution caused by Lorentz Barrel and Drum. Lorentz Barrel, a San Jose company, is a 40-year-old drum-recycling business that collects old drums, washes them out, repaints and sells them.

In many instances, the drums have contained hazardous wastes that have contaminated the soil and groundwater at the site. The company's owner was sentenced to 2 years jail and fines and penalties of US$2.04 million for violation of the state hazardous waste laws, but he died before his imprisonment could take place.

The Lorentz site had been on the California Superfund list of the state's most hazardous waste sites. Soil at the site and shallow groundwater directly beneath the site were contaminated with benzene, 1,2-dichloroethane, trichloroethylene, and vinyl chloride (Table 4.2).

Although contamination of the deeper aquifers that supply drinking water to residents has not been detected, the EPA is nevertheless concerned about the potential of contaminants to affect these aquifers in the future. It will take much time and money to clean up the site. The investigation has already cost between US$5 million and 15 million. At the community meeting, citizens asked various questions about the effect of pollution on the deep well, air pollution, and the responsibility of the original owner of the drums. The citizens movement is trying to put pressure on officials to:

1. Get rid of all barrels at the Lorentz site;
2. Determine exactly how far chemicals have spread underground;
3. Identify and seek reimbursement from the companies that originally sent their waste drums and chemicals to Lorentz's company;
4. Monitor air pollution at the site and within a 1-mile radius of the site; and
5. Conduct an immediate health survey of residents and workers.

To effect the environmental cleanup, the local government is trying to release information, gather information from residents living in close proximity to the company, answer their questions, and satisfy their requirements. This tradition of "grass-roots democracy" represents a good guide for environmental policies, especially in Japan. I was impressed that the community should hold such a meeting, and that, after heated discussion, the participants confirmed a decision to achieve a "well well".

After carrying out emergency treatment of contaminated surface soil, State Health officials have decided to turn over responsibility for cleaning up the site to the USA EPA. There are two key reasons for this decision:

1. The state has failed to hire clean-up contractors who meet the EPA's standards; and
2. The state needs to turn over complicated and expensive cleanup jobs like that at the Lorentz site to the federal Superfund hazardous waste program.

In March 1988, the EPA and California Department of Health Services completed removal of 1000 drums of hazardous materials. The EPA plans to cleanup contaminated groundwater. The water will be stripped of chemicals and dumped into nearby Coyote Creek, which flows into San Francisco Bay. A list of potential responsible parties, that is, the companies—including Hewlett Packard and IBM—that sent their waste drums to Lorentz's company has now been drawn up.

In California, there are 40–50 drum recyclers, many of whom run small-scale family businesses. Six drum recycling businesses in California are on the State Superfund list of California's worst hazardous waste sites, and, in many instances, the owners cannot afford to pay for the cleanup. State environmental officials estimate that they will need to spend between US$10.5 million and 20.5 million to clean up the six state Superfund sites (*San Jose Mercury News*, March 14, 1988).

Because drum recyclers take over the responsibility of the treatment of chemical waste instead of the chemical manufacturers and semiconductor industries themselves, such an assumption of responsibility by contractors should be defined as a "shuffling off" of the duty of waste disposal by the industries concerned.

2.3 The Danger of Silicon Valley

2.3.1 Toxic Gas Release

Ambulances and fire fighters are often called out not only to deal with traffic accidents, but also to extinguish vehicular fires and container explosions. In Silicon Valley, in particular, we can observe the transport of toxic gas on public routes. The semiconductor industries bring in their gases from outside. In California, during the period from January 1984 through to December 1987, a total of 212 hazardous material shipment incidents were reported, and 258 persons were directly exposed to the commodity involved—mainly corrosive and liquid materials.

According to *Semiconductor Industry Study* by Cal/OSHA (State of California, Division of Occupational Safety and Health 1981[13]), in 1979, 42 semiconductor companies in Silicon Valley used 570000 gallons of solvent, 1.56 million gallons of acid, and 1.55 million cubic feet of gas. A city-by-city report (*Toxic Release Inventory Form*) for 1992 says that a total of 8.5 million pounds of toxic chemicals have been released within Silicon Valley. This is an extremely high concentration of toxic chemicals. In Silicon Valley, many toxic gas accidents have already taken place. For example, silane (SiH_4), which is most frequently used by the semiconductor industry as a carrier gas, is potentially explosive. In March 1988, in New Jersey, a silane gas explosion killed three people. A mixture of silane and laughing gas (nitrous oxide) seemed to be the cause of the explosion, whereas in Silicon Valley itself, local people near route 101 were evacuated during an investigation into the escape of silane gas made by the same manufacturers involved in the incident in New Jersey.

Faced with incidents such as these, Professor Kenneth Mackay, a meteorologist at San Jose State University, has carried out a series of toxic gas release experiments. Using the EPA's *Chemical Emergency Preparedness Program* (1985) as a guideline, he investigated four toxic gases (arsine, chlorine, diborane, and phosphine). Arsine is a highly poisonous, inflammable gas that destroys red blood cells. The report found that five companies in the electronics valley 50 miles away stored enough arsine to endanger the health of people within a 1.5-mile radius if they were to breath the gas for several hours in the event of a major release of arsine into the atmosphere.

The five companies listed in the report are Raytheon in Mountain View, Advanced Micro Devices in Sunnyvale, Exel in San Jose, and Precision Monoliths and Epitaxy, both in Santa Clara. This raised the possibility of a major "catastrophe", the "death of the valley" and "a second Bhopal" in Silicon Valley. Bhopal is a city in India, where, in December 1984, a pesticide factory belonging to Union Carbide, a USA multinational company, exploded and killed more than 2500 people.

Such fears are not unfounded because, in the USA between 1974 and 1986, there were 69 cases of fire in semiconductor factories, 28 of which were caused by the ignition of flammable or pyrophoric combustion gases (hydrogen, silane).

After 5 years discussion, the nation's first toxic gas model ordinance has finally been passed by each city in Silicon Valley. The new law requires the secondary containment of all toxic gases, neutralization apparatus, and monitoring equipment. More than 200 businesses will need to spend US$100 million to comply with the law according to the Santa Clara County Manufacturing Group.

2.3.2 1989 Earthquake and High-Tech Industry

During the earthquake of October 17, 1989, chemicals were spilt onto the floor of two Silicon Valley electronics firms—the FMC Corporation plant in San Jose and the National Semiconductor plant in Santa Clara. Silicon Valley, however, was spared a catastrophic chemical accident. According to Bay Area fire fighters, the avoidance of major accidents was due to the tough hazardous material laws. Moreover, the new toxic gas ordinance requires a seismically activated valve that will shut off the gases in the event of an earthquake.

2.3.3 Air Pollution in Silicon Valley

In Silicon Valley, which lies at a drift point on San Francisco Bay, business starts early; some factories begin work at 6 a.m. This is to avoid traffic jams. From the foot of the mountain, smog spreads across the highways.

An EPA study reported, in May 1987, that ailments caused by dirty and dusty air kill as many 57 residents of Santa Clara County each year. According to the EPA, the main polluter is the motor vehicle. Other organic gas emitters are degreasers for metal processing, industrial solvent coatings, and photoresists used in semiconductors. However, air pollution caused by the semiconductor industry is still not being thoroughly investigated. According to a monitoring survey carried out by the Bay Area Air Quality Management District, precursor organic compounds from the semiconductor industry (mainly from solvent sinks, mix stations, and photoresist developers) total 11 000 pounds per day; however, this type of monitoring is not yet able to detect the presence of inorganic toxic gas. We need much more information about air pollution caused by the semiconductor industry. *A Survey of the IC Industry and Environmental Protection* (1987) by four Japanese ministries said: "Untreated trichloroethylene and tetrachloroethylene have been detected in the air near the emission apparatus. More attention should be paid to this issue."[14]

In August 1990, Bay Area Air Quality Management District released a "Toxic Hot Spots" list of significant air polluters. The new list of 123 facilities around the Bay Area includes several of Silicon Valley's most prominent electronics firms—IBM, Lockheed, and United Technologies, National Semiconductor. The Clean Air Act Amendments of 1990 lists 189 toxic pollutants. Factories must install "maximum achievable control technology" to reduce the release of these pollutants by 90% by the year 2000.

In place of troublesome organic solvents, many semiconductor factories began to use CFC (freon gas) as a cleaning agent. In Silicon Valley, IBM's San Jose plant purchased more than 2.6 million pounds of CFC in 1987 and released nearly 1.5 million pounds of CFC in 1987. That probably makes IBM the biggest high-tech user of CFC in Silicon Valley (*San Jose Mercury News*, July 11, 1988). Yet, even a cleaning agent like the CFCs has a negative effect on the environment; CFCs are now known to be destroying the ozone layer and causing the Greenhouse Effect. Although IBM claimed that air emissions of CFC-113 had been cut by 25% in 1988, they would still have ranked as the number one polluter in the state, because their emissions were greater than that of any other industrial plant. The company committed itself to a 1993 deadline to phase out CFC usage altogether and accomplished this objective. In section 4, I shall analyze CFCs in more detail.

2.3.4 Pollution and Safety Problems Relating to Military Facilities

When I drive along route 101, the artery of Silicon Valley, I can see a big hangar beside San Francisco Bay. This is the Naval Air Station, Moffett Field (base camp for the P3 anti-submarine patrol plane). Next to Moffett are located the Lockheed Missiles and Space Company and the Onizuka Air Force Base, one of the top five American military command and control centers. Therefore, there are military aspects to the Silicon Valley problem.

The USA nuclear weapons industry is currently in a state of crisis, rocked by leakage of radioactive waste, equipment breakdowns, and plant shutdowns. The groundwater around many facilities is contaminated by toxic and radioactive waste. The contaminants include plutonium, cesium, PCB, chromium, and solvents. According to the Department of Energy, it will cost US$66 000–110 000 million to clean up the mess.

Many of the military base camps surrounding San Francisco Bay store nuclear weapons and are also contaminated by chemicals. Among them, Moffett is the worst and is listed as a federal Superfund site. Nineteen separate sites on the facility have been contaminated with millions of gallons of toxic waste over the past 40 years. The contaminants include petroleum hydrocarbons, solvent, oils, metals, paints, and battery acid.

In July 1988, a community meeting was held near Moffett Field. Besides Moffett, Intel and NEC are the main polluters. Some retired couples attended this meeting. Because they were worried about the drinking water, they discussed the safety of boiled water and how to press officials to take proper preventative action.

Missiles also present safety problems. We easily remember the explosion of the Space Shuttle, Challenger. In the USA, between 1987 and 1988, there was more than one explosion of a missile fuel facility: in December 1987, at the fuel factory for MX missiles in Utah; in May 1988, at the fuel plant for the Space Shuttle and the Titan missile in Nevada.

United Technologies, Chemical Systems Division, southeast of San Jose, handles the same fuel as the Nevada plant and also manufactures missiles such as the Minuteman, Tomahawk, Trident, and Titan. The Air Force inspectors found defects in an average of one of five components for missile motors. The firm was cited for "severe fire threats" and explosives hazards (*San Jose Mercury News*, November 24, 1987). The company's engine tests and open-air burning of excess rocket fuel impacts neighbors up to 20 miles away and spews forth almost one million pounds of highly toxic chemicals into the air, including hydrogen chloride and aluminum oxide.

Recently, soil and groundwater contamination has occurred at this site owing to leakage from three hazardous waste ponds where various hazardous waste products were stored. At the same time, organic solvent was seeping through the ground close to Anderson Reservoir, which provides drinking water for up to 300000 people. In August 1988 and September 1989, the Altus Corporation, San Jose, which manufactures lithium batteries that are sold primarily to the military, belched huge clouds of black toxic smoke. This plant has been cited by federal and state agencies for 15 violations during the past 5 years. This dangerous factory is located in the center of San Jose City.

2.4 Occupational Safety Problems

2.4.1 Health Hazards to Semiconductor Industry Workers[15]

Debbie Berry suffered severe headaches and difficulty in breathing. Between 1985 and 1986, she worked for a semiconductor company, Siliconix Santa Clara, engaged in cleaning with cellosolve (ethylene glycol monoethyl ether). Because she suffered from a stroke and trembling hands, she was fired. She claims that her work would have been safer had there been a correct ventilation system and alternative chemicals. Approximately 200 persons are suing the company: "We are guinea pigs".

After working for several years for the semiconductor company Signetics, Gene Lemon, a chemical engineer, developed a disease of the respiratory organs as a result of exposure to chemicals; he sued the company. Therefore, in Silicon Valley, semiconductor workers and engineers face many health hazards. During the 1970s, nausea and headaches suffered by semiconductor workers in Silicon Valley attracted attention.

Dr. Joseph LaDou,[16] the author of "The not-so-clean business of making chips" (*Technology Review*, May/June 1984), has tackled this type of problem. He is Clinical Professor of Occupational and Environmental Medicine, University of California, San Francisco, and also works at the Peninsula Industrial Medical Clinic, Sunnyvale.

TABLE 4.3. Occupational illnesses in the semiconductor industry (California Workers' Compensation Statistics)

	1986	1987	1988	1989	1990	1991
Occupational Illness as Percent of Work-loss Cases						
All Manufacture	7.0	7.0	7.5	6.9	8.4	9.1
Electronic component	16.9	18.7	17.3	14.4	19.9	16.5
Semiconductor	21.1	26.4	22.4	16.5	21.1	18.4
Systemic Poisoning as Percent of Occupational Illness						
All Manufacture	20.6	17.4	17.0	16.6	15.6	13.9
Electronic component	39.7	31.4	30.7	35.6	28.5	28.7
Semiconductor	49.6	34.3	39.8	40.8	37.9	25.6
Systemic Poisoning as Percent of Work-loss Cases						
All Manufacture	1.5	1.2	1.3	1.1	1.3	1.3
Electronic component	6.7	5.9	5.3	5.1	5.7	4.7
Semiconductor	10.5	9.1	8.9	6.7	8.0	4.7

According to Dr. LaDou, approximately 50% of the injuries and illnesses suffered by workers in the semiconductor industry are not reported to the federal survey.

Furthermore, until 1981, accidents caused by exposure to toxic gasses had been reported as "illness"; after 1981, however, these accidents were reported as "injuries". Therefore, incidents of gas exposure are only reported in serious cases.

On the other hand, the California Workers' Compensation Statistics has collected more detailed data about the semiconductor industry in California. Statistics for Occupational Illness as Percent of Work-loss Cases, Systemic Poisoning as Percent of Occupational Illness, and Systemic Poisoning as Percent of Work-loss Cases reveal that the number of such cases in the semiconductor industry is relatively high, sevenfold higher than in any other branch of manufacturing (Table 4.3).

Even the Bureau of Labor Statistics (1984, 1985) admitted that the percentage of illness attributed to "respiratory conditions due to toxic agents" and "skin diseases or disorders" was higher for the semiconductor industry than for manufacturing overall. This relatively high rate of systemic poisoning in the semiconductor industry is of note.

There are several sources of potential danger: (1) many sorts of chemicals and gases; (2) non-ionized radiation (laser, ultraviolet); and (3) ionized radiation (α-, β-, X-ray). There are various sorts of health hazards here: (1) carcinogenic (arsine, benzene, and trichloroethylene); (2) reproductive (cellosolves, lead); and (3) immunologic (so-called "chemical AIDS"). Workers fall sick after acute exposure, or long-term, low-density exposure. Dr. LaDou states that because high-speed microelectronic devices often require wafers composed of arsenic and gallium, these poisons are being used in ever-larger quantities. As a result, the production of gallium arsenide wafers requires much greater care if the workers are not to be endangered by the toxic arsenic powder. At the same time, however, safer materials should be, and now are being, developed, and the old materials are being replaced.[17,18]

As the *Semiconductor Industry Study* (1981) published by Cal/OSHA[19] stresses, "unusual" work, such as maintenance, can also cause trouble. In Japan, from 1966 to 1985, there were 22 cases of toxic gas exposure during this sort of work in the semiconductor industry. However, *A Survey of the IC Industry and Environmental Protection* (1987) by four Japanese ministries said, "No information is available about the work of changing gas cylinders, pump oil, or maintenance work such as piping."[20] Further studies need to be undertaken and improvements should be carried out.

Nor should the safety problems of minority workers at high-technology related companies be ignored. For example, female minority workers often use organic solvents at "garage"-size work places that manufacture printed circuit boards. As the cost of living—especially housing—has climbed rapidly in recent years, so too has the proportion of Asians living in Silicon Valley, because these people seem more willing to share tiny living spaces. At the circuit-board manufacturer Flextronics, where unskilled jobs are plentiful, roughly 50% of the work force has an Asian background, while 20% are Hispanic (*San Francisco Chronicle*, July 18, 1988).

Early in the morning, many people stand in line at the federal emigration office in San Jose. Silicon Valley, a center for high-technology, is, in fact, also a center for refugees from Vietnam. Approximately 100 000 Vietnamese live in San Jose. Light railroad notices are written in English, Spanish, and Vietnamese. Some of the Vietnamese work in the electronics industry. However, the real state of things is not clear.

At the same time, as more multinational companies in the semiconductor industry are being located in southeast Asia, the occupational exposure of workers has begun there too. For example, in January 1983, at an electronics factory making DC motors in Hong Kong, owned by Mabuchi Motors of Japan, 193 female workers were sent to hospital after breathing high levels of ozone, phosgene, and other gases that had been slowly released by printing equipment that used ultraviolet light. Thirteen of the victims were pregnant at the time of the incident. According to some of the Mabuchi workers, at least six of those women experienced miscarriages or underwent forced abortions because of fetal death.

By 1992, Malaysia had become the largest exporter of semiconductors in the world. Vivian Lin carried out research of the working conditions of five semiconductor facilities (assembly line) in Malaysia and Singapore, where night work by women is allowed. Her survey shows that workers using solvents regularly are more likely to experience problems with menstruation, pregnancy, and childbirth.[21]

2.4.2 Spontaneous Abortion among Semiconductor Workers

In January 1987, AT and T, a well-known telecommunications company and also a large semiconductor manufacturer, banned pregnant women from production lines because of concern about employees' exposure to chemicals that might cause miscarriages. AT and T implemented this policy after learning about a health study in Massachusetts that reported an increased rate of miscarriages among chip production workers.

The Division of Public Health, University of Massachusetts, carried out this health study of 744 employees at DEC, Hudson LSI. Personal interviews were conducted with manufacturing workers, the spouses of male workers, and an internal comparison group of non-manufacturing workers. Elevated spontaneous abortion ratios were observed for females working in the diffusion (38.9%) and photolithographic (31.1%) processes.[22-24]

Various general symptoms of sickness, such as arthritis, nausea, rashes, sore throat, and headache, were examined and were reported more frequently among workers in the manufacturing process than among non-exposed workers.

Because approximately 66% of workers in the semiconductor industry are women in 1984, the Massachusetts study has revealed a serious problem. Critics have attacked the method of this study, and the federal OSHA admits that it needs to be followed up by much more investigation. AT and T, IBM, DEC, and SIA have carried out more research. Based on the IBM and SIA studies, these companies will phase out the use of glycol ether (EGE) at their plants. DEC and AMD allow a pregnant woman the option of transferring to another job. The response of the Santa Clara Center for Occupational Safety and Health, an organization seeking workers' safety, is "Remove chemicals, not workers". The Santa Clara Center for Occupational Safety and Health claim that AT and T has adapted a discriminatory policy of excluding pregnant women from certain production areas, without considering the effect of toxic chemicals on the reproductive systems not only of women, but also of men prior to conception. Policies singling out pregnant or fertile women for removal may violate Title 7 of the 1964 Civil Rights Act, which forbids employment discrimination on the basis of sex or pregnancy. As a result of exclusive policies, women have lost employment opportunities. Companies are entitled to protect themselves against the possibility of lawsuits brought by the injured offspring of employees. Workers argue that companies should: (1) replace substances known or suspected of causing harm with safer substitutes; (2) institute engineering controls, such as enclosure or ventilation; (3) transfer workers temporarily without loss of wages, seniority, or benefits, while efforts proceed to reduce the hazard; (4) finance independent industry wide studies to assess the health effects of workplace exposures; (5) integrate the concept of occupational and environmental health and safety into process design; and (6) provide independent training and positions for workers, and health and safety representatives.

California State used to require its own OSHA (Division of Occupational Safety and Health) to inspect and regulate over a wider range than the federal OSHA. In 1987, however, Governor George Deukmejian abolished the Cal/OSHA for financial reasons, notwithstanding that the annual cost of Cal/OSHA was really only US$6.8 million–8.4 million, less US$1.6 million in loss and penalties. Such expenses are not great when we consider that the overall size of the state budget is US$42 billion. However, there has been wide private criticism among businesses. The reason is simple: Californian employers are already burdened by one of the most expensive compensation systems for job-related illnesses and injuries (*San Jose Mercury News*, October 22, 1987). After the abolition of the Cal/OSHA, inspections dramatically decreased, and workers deaths increased by approximately 50%. Therefore, in November 1988, a state ballot voted to restore the Cal/OSHA.

This section has demonstrated that Silicon Valley not only suffers from groundwater contamination by organic solvents, but also faces many other problems that affect the environment, general safety, and occupational health.

3. Tackling the Cleanup

3.1 The Environmental Protection Movement in the USA

3.1.1 Citizens Cleanup

Although Ted Smith is an attorney, he no longer stands at the bar. He is engaged, instead, as the Executive Director of Silicon Valley Toxics Coalition. Originally, as a lawyer who specialized in labor problems, and with his wife, Amanda Hawes, who specialized in workers' safety, he tackled the Fairchild case. The coalition is composed of many sorts of citizen groups, environmental groups ("Citizens for Better Environment"), occupational safety groups, and labor unions (AFL-CIO). Because unions are not permitted to organize workers at high-tech industries, they make much of occupational safety and environmental problems. Consequently, the high-tech industries themselves regard the Coalition as a separate arm of the unions.

The American environmental protection movement has two main divisions: one is the wildlife-protection movement, like the Sierra Club; the other is the anti-toxic waste movement. The Silicon Valley Toxics Coalition, however, has a good relation with other wildlife-protection movements, while the environmental movement is building a closer connection with the labor movement as a result of tackling nuclear waste and toxic waste problems.

The first project undertaken by the Silicon Valley Toxics Coalition has been to fight for local legislation to require safer containment and monitoring of chemical storage, and to require public disclosure of toxic chemicals stored in the communities: (1) to persuade the EPA to declare the Santa Clara Valley sites Superfund sites; (2) to ask the state to carry out a study of birth defects in the Fairchild spill area: (3) to ask the County Board of Supervisors to establish a Safe Water Council to move on solutions to toxic waste problems; (4) to push the Legislature to appropriate more funds for toxic hazard identification and cleanup; (5) to ensure safer toxic chemical storage practices; and (6) to encourage government and industry to move faster on cleanup efforts.

Hereafter, I would like to analyze how far these objects have been achieved.

3.1.2 Proposition 65[25]

It is noteworthy that California voters overwhelmingly (68%) adopted Proposition 65, the Safe Drinking Water and Toxic Enforcement Act of November 1986. Governmental agencies will no longer consider the use of chemicals known to cause cancer or reproductive toxicity "innocent" until proven "guilty" of harming public health. After February 1988, anyone who, in the course of business, exposes any individual to a chemical "... known to the state to cause cancer or reproductive toxicity" must first provide "clear and reasonable" warning of the exposure. This new regulatory process creates incentives for businesses to cooperate with governmental agencies in establishing levels that define "no significant risk" and to reduce their use of toxic chemicals.

Most of the high-tech industries naturally opposed Proposition 65, and gathered US$0.6 million to organize an anti-proposition campaign. On the other hand, Hollywood actors and actresses supported the citizens' campaign for Proposition 65.

In Californian restaurants, you are asked to observe that there are smoking or non-smoking areas for diners. Official warnings about smoking are already familiar to the Japanese. Similar warnings about chemicals that could cause birth defects or other kinds of reproductive harm are printed on labels or are available on free-phone. The focus of the problem is how to decide what constitutes "significant risk". For example, the carcinogenic potential of alcohol, charcoal-cooked steak, and furniture polish etc., has also been considered.

In February 1987, the Governor of California named a scientific advisory panel of 12 scientists to advise him about which chemicals should be included on the Proposition 65 list of carcinogens and reproductive toxicants. As of 1990, the list has grown to include about 350 compounds.

We have to take note that power of Proposition 65 to prevent environmental pollution is limited by its inability to control the substantial risks associated with high background levels of carcinogenic exposure. At the same time, "no significant risk" ignores the potential magnitude of cumulative cancer risks from multiple exposures to different chemicals. In establishing its "significant risk" standards—one additional case of cancer per population of 100000 (which means 280 persons in California State)—the administration split the difference between the business and environmentalist positions. Environmental groups had asked that one additional cancer per 1 million people be the standard of unacceptability, while businesses asked that it be one per 10000.

Although businesses and the governor fear that Proposition 65 could frighten people unnecessarily and have a negative economic effect, we have to remember that the background of Proposition 65 is that people already genuinely fear environmental pollution.[26]

According to a telephone survey of 1000 Santa Clara County residents (June 1988) about water quality, 68% of people believed that it was a "big problem", while 54% thought that ". . . there are probably some unsafe things in the water that you cannot see". In fact, because approximately half the Santa Clara County residents no longer drink tap water, bottled mineral water sells very well.

Because Proposition 128 of 1990, known as "Big Green", which was trying to phase out 20 cancer-causing pesticides, to ban CFC, to protect the coast, to reduce emissions of global warming gases by 20%, and to protect old redwoods, was not passed, Proposition 65 grows ever more significant.

3.2 Conduct of Government

3.2.1 EPA (US Environmental Protection Agency)

"Political Football" is a term coined by Lenny Siegel, a coauthor of *The High Cost of High Tech*,[27] about the response of government to pollution in Silicon Valley. As he points out, high-tech pollution is kicked about between federal, state, county, city, and district.

The EPA, the agency responsible for the whole environment program and money procurement, has made much of the pollution of Santa Clara County, and has proposed to establish an "Integrated Environmental Management Project" (one of four nationwide projects) for this area.

This project's goals are: (1) to evaluate and compare the health risks from toxic pollutants in the environment; (2) to use these pollutants in the evaluation to set informed priorities for further analysis and possible control; and (3) to work closely with government agencies and the community to manage environmental public health problems effectively.

In October 1984, the EPA listed 19 Superfund sites in the Santa Clara County area. Since then, the number of Superfund sites has grown to approximately 33, 10 of which have been transferred to another program. Superfund was to be funded from taxes on crude oil and 42 different commercial chemicals. State governments were to pay 10% of the cost of Superfund work at privately owned sites and 50% at those that were publicly owned. Superfund was to be used in cases where responsible parties could not be held accountable or were unable to respond to emergency situations involving hazardous substances. There would be some cases of "mixed funding" between the fund and responsible parties to share remedial costs. In Santa Clara County, however, businesses do not always support the Superfund list.

3.2.2 California Department of Health Services (DOHS)

The California DOHS judges water quality, manages large-scale wells, and is responsible for the application of the State Superfund. This department has carried out epidemiological studies of Los Paseos. The action level for the state, that is, the regulation level of water insisted on by California State, is more strict than that of the World Health Organization (WHO) and Japan. The regulation levels set by the WHO and Japan are: tetrachloroethylene 10 ppb, trichloroethylene 30 ppb, and 1,1,1-trichloroethane 300 ppb. The comparable action levels of California State are 4, 5, and 200 ppb, respectively. Susanne Wilson, chair of the Safe Water Council of Santa Clara County, has, nonetheless, criticized the policy of the state. She recommended that the regional water board promptly receive an augmentation of 22 staff members. The State Department of Health Services has rejected this request. As long as the contamination remains below state action levels, the state has defined all possibilities of additional state purification as irrelevant. Ms. Wilson's great concern is that individual firms or state agencies may view the application of state and federal water quality standards as a license to pollute. She added that the delay in making use of Superfund is the fault of the state. In 1985, when the State Assembly proposed a US$8.5 million cleanup program, the Governor rejected it. In 1991, a part of DOHS was reorganized into the Cal/EPA.

3.2.3 Regional Water Quality Control Board (RWQCB)

The California RWQCB, San Francisco Bay Region, has approximately 100 members of staff, and is responsible for managing and investigating water quality and directing each company. The State of California has nine Regional Control Boards, all of which are short of staff. The local government has direct responsibility, but few resources or means.

3.2.4 Santa Clara County Public Health Division (SCCPHD)

The County Authority is a branch of local government between the state and the cities, responsible for fire, and public health, welfare, and safety. The SCCPHD controls

TABLE 4.4. Hazardous materials program in California

Hazardous Materials Control
1. State Regulation of Underground Storage Tanks (Sher Bill)
2. Hazardous Materials Storage and Emergency Response (Waters Bill)
3. Acutely Hazardous Materials Risk Management (La Follette Bill)
4. Emergency Planning and Community Right-to-Know (SARA Title III)
5. Uniform Fire Code (1988 Edition)
6. Worker Right-to-Know (Federal OSHA, Cal/OSHA, Occupational Carcinogens Control Act)

Hazardous Waste Regulation
1. Hazardous Waste Management Planning and Facility Siting (Tanner Bill)
2. Federal Regulation of Hazardous Wastes (RCRA, HSWA)
3. Federal Toxic Cleanup Law (CERCLA, SARA)
4. State Superfund, Hazardous Substances Account Act
5. Toxic Pits Cleanup Act (Katz Bill)
6. State Regulation of Hazardous Solid Waste Disposal (Calderon Bill, Eastin Bill)
7. State Hazardous Waste Control Law (HWCL)
8. Hazardous Waste Management Hierarchy
9. Incentives and Assistance to Improve Hazardous Waste Management
10. Household Hazardous Waste (HHW) Management

Other Toxics Laws and Programs
1. Proposition 65 (Safe Drinking Water and Toxic Enforcement Act 1986)
2. Asbestos (OSHA, EPA), (DHS, Cal/OSHA)
3. Safe Drinking Water (Federal SDWA, California SDWA)
4. Industrial Waste Pretreatment Programs, (Controls Toxic discharge into Sewers)
5. Clean Air Act (Federal CAA, Cal CAA, "Toxic Hot Spots" Bill)
6. Water Quality Control (Federal Clean Water Act, Porter-Cologne Water Quality Control Act)
7. Pesticides (FIFRA, Cal Pesticide Contamination Prevention Act, Restricted Materials Act)
8. Transportation of Hazardous Materials and Wastes (Fed Haz Mat Trans Act; RCRA, Cal Vehicle Codes & C 8)

Source: EPICS International (1990)

small-scale wells and regulates underground storage tanks. This Division has approximately 10 toxicologists, who must investigate the possible contamination of approximately 1200 private wells. Although, in 1985, a special committee, composed of members from each branch of local government, was set up, the real responsibility for pollution control is not clear. Santa Clara County has approximately 24 federal and state pollution control laws (see Table 4.4), but each law is very complicated and some are duplicated.

3.2.5 Santa Clara Valley Water District (SCVWD)

Before Santa Clara County became a concentrated high-tech area, it had many orchards, and, approximately 70 years ago, this Water District was authorized to control agricultural water and flooding. Today, the District gathers the groundwater tax and collects data on groundwater quality; it is also responsible for closing contaminated wells.

In January 1982, after the Fairchild case, the Santa Clara Fire Chiefs Association proposed a Hazardous Materials Model Code that would provide double containment

and strict monitoring of all toxic chemicals. In 1983, the County and each city implemented Model Ordinances based on this proposition. Later, state laws were passed
to regulate toxic chemicals. Virtually all the 4000 facilities that have submitted
Hazardous Materials Management Plans have been inspected. Approximately 500
double-containment tanks have been constructed, 1400 tanks have been transferred,
and 322 tank leakages have been reported. The ordinance has, therefore, taken effect,
to some extent, specifying and examining hazardous materials storage sites.

3.2.6 City

Each city has implemented a hazardous materials storage ordinance to regulate underground storage tanks and to control toxic chemicals. With regard to the "trade secrets"
relating to the chemicals used by the industries, San Jose City Hazardous Materials
Storage Permit Ordinance, "Trade Secret", for example, stipulated that, "The exact
name of the trade secret material be placed in a double-keyed lock box and maintained in at least two locations at the facility, key shall be accessible to permittee's designated emergency response person." I have heard that, at the time of implementation,
discussions focused on secret policies against terrorists and rival Japanese companies. Today, San Jose City is providing guidelines for user financing, information
advice, and treatment of household toxic waste.

3.2.7 Integrated Environmental Management Project

The purpose of the Integrated Environmental Management Project is, as I have said,
to identify and define risk to public health posed by exposure to toxic contaminants,
such as polluted groundwater, and air trihalomethanes (THMs), to assess the relative
risks of exposure to such dangers, and to control the likelihood of such risks more
efficiently.

The Draft Stage One Report released in October 1985, focused on 30 contaminants,
polluted air, and the carcinogenic potential of polluted drinking water. On the one
hand, this report stressed the risks of air pollution and the presence of THMs in
surface water. On the other hand, the report estimated only low risk from contaminated groundwater. The Silicon Valley Toxics Coalition criticized the report because
it concentrated only on 30 substances and excluded the risks from stored chemicals
during fires or earthquakes. The risk assessments tend to be manipulated.

The Stage Two Report appeared in September 1987; this report had carried out a
cost–benefit analysis of the cleanup of groundwater and surface water. The report
stated that ". . . widespread treatment (of groundwater) may be more costly than
obtaining replacement supplies. Also, closure (of well) is projected to entail a greater
risk than treatment, since users of closed wells are assumed to be provided
with surface water replacement supplies (which present a risk of cancer from
THMs)."[28]

After discussions between government, business, and citizens, an action plan for
this project agreed over the air, water, and institutional issues. Since I listened to the
discussion, I would like to mention some of the controversial issues raised. Business
asserted that their efforts for cleanup should be evaluated favorably, but that the technological problems were large and complicated. Although business hoped that clean

air regulations applied to special industries would not be necessary, the "action plan" agreed on the need to ". . . improve its data base on emissions from industries, including semiconductor and high-tech facilities" (item 5).

Although the labor side asked for the outright banning of cellosolve (glycol ethers), it was decided that ". . . further research leading to phasing out the use of toxic glycol ethers should be performed" (item 6).

Water issues were dealt with accordingly: (1) as to the hazardous materials ordinance, "Coordinate inspection activities to minimize duplication", and "Explore the possibility of centralizing data collection"; (2) as to private wells, "complete the private wells testing program", and "implement the following testing"; and (3) as to aquifer management, contrary to the opinion of business, "Where restoration of the aquifer is too costly, responsible parties fund measures that mitigate or compensate for the residual impacts of their contamination" (item 14), while "Protection zones should be developed" (item 17). On the other hand, the Silicon Valley Toxics Coalition proposed a "Waste Reduction" scheme.

The action plan evaluated the status quo over institutional issues; ". . . there is no formalized, well-defined process for toxics management decision-making". Therefore, the action plan proposed the establishment of a Santa Clara County Toxics Policy Council. This Council was to be characterized as a joint powers authority and be based on multiparty agreement.

Although this Council has not been set up, the Tanner Advisory Committee attached to the County Supervisor is now working to improve the management of toxics and solid waste.

It is noteworthy that, in order to decide upon an environmental policy, the government proposed a draft plan openly and organizes discussion and full negotiation. Japanese environmental policy making should follow this example.

3.3 Tackling Cleanup by Businesses

3.3.1 Fairchild

When I visited the Fairchild Company, San Jose, only the aeration machine was at work. Well #13, 2000 ft away, had been closed down in 1982. After discovery of the pollution, Fairchild removed 3389 cubic yards of contaminated soil, and extracted and treated groundwater to reduce the level of chemical concentration (the 1,1,1-Trichloroethane level was 24 ppb in January 1990). Fairchild also installed an underground slurry wall, which is 3 ft wide and runs vertically to depths from 80 to 140 ft beneath the plant site. However, according to the EPA, the slurry wall is not the final remedy and the citizens point out that slurry walls have a tendency to leak. In September 1987, the National Semiconductor Company bought out the Fairchild Company. Fairchild, however, retained all environmental liabilities associated with its past activities at the site. The San Jose Fairchild site is to become a shopping center.

3.3.2 IBM

IBM is the largest captive semiconductor manufacturer in the world. IBM, San Jose, has 14200 employees, the third biggest company in Silicon Valley (the largest is

Lockheed Missile and Space, and the second largest is Hewlett Packard). The IBM plant is located approximately 1.5 miles northwest of the Fairchild site. Some IBM employees live nearby. Because few Fairchild employees live in Los Paseos, local hatred of Fairchild has been very strong. On the other hand, IBM has a "company town", where a citizens movement is difficult to organize. In fact, an IBM employee was originally involved in the Fairchild case, but later dropped out.

Contamination from the IBM plant has already affected approximately 25 public and private wells. More than 100000 San Jose residents are potentially exposed to contaminants in these wells. IBM has already spent US$200 million in cleanup and dug more than 300 monitoring wells and 19 pumping-up wells. Nonetheless, the Silicon Valley Toxics Coalition criticized the methods of this cleanup.

1. It is necessary to dig more monitoring wells and to grasp the extent of the pollution more exactly. Since the RWQCB ordered IBM in 1984 to clean up only the "first area", there is the possibility of pollution in the lower northern area.
2. IBM is pumping up to 12 million gallons of contaminated groundwater into Canoas Creek every day. However, this water could easily pollute other groundwater.
3. The aeration used for cleanup may cause air pollution.

Because of the Californian water shortage, IBM reused 68% of the water it pumped in August 1990 for irrigation or for on-site manufacturing purposes. IBM has a toxics division and a stricter environmental standard than the federal government. It will be interesting to see how IBM tackles further cleanup in San Jose.

3.3.3 Clean Water Task Force

The cost of cleanup is higher than the land price itself. For example, in the case of Fairchild, the land price was US$5 million, whereas the cleanup cost US$30 million.

The Comprehensive Environmental Response Compensation, and Liability Act of 1980 (CERCLA), known as Superfund, defines four categories of persons who are financially responsible for hazardous waste cleanup: past owners, present owners, transporters of hazardous substances, and generators of hazardous substances. As a result, current property owners who have not caused contamination may still be financially liable for any contamination associated with their land. This concept may extend to financial institutions. The Superfund Amendments and Reauthorization Act of 1986 (SARA) amended CERCLA and created an "innocent purchaser" defense to owner responsibility. To claim innocence, however, the landowner must first demonstrate that, at the time of property acquisition, there was "no reason to know" that the property was contaminated. Because investigation of pollution is, therefore, necessary, the expert is naturally eagerly sought after.

The Clean Water Task Force was consequently formed in 1984 to encourage and support activities by member industries in Santa Clara Valley to protect and conserve drinking water. It is an organization sponsored by the Santa Clara County Manufacturing Group. The environmental Programs Director of the Clean Water Task Force of the Group, Jacqueline Bogard, is another "iron lady", confronted by environmentalists. According to a survey of 74 companies, since 1982, 254 extraction wells have been installed and 15 million gallons per day have been extracted. The water reuse rate is only 1.11%. A total of 2006 monitoring wells are actively being sampled. More than US$65 million has already been spent on investigative studies at these sites. The number of secondary containment tank systems rose from 211 tanks in 1982 to 372 tanks in 1987. The companies surveyed have spent more than US$28 million since 1982 on preventative measures. Because the RWQCB requires industry to sample many monitoring wells and because this takes much time and money, industry has expended enormous resources on investigative work.

According to Jacqueline Bogard, the EPA's Superfund is not necessary, because, in Silicon Valley, industries themselves have already taken action to clean up pollution. In the case of Superfund, the EPA requires much documentation and the federal fund cannot be used except in cases of bankruptcy. "How is it possible for society to use limited resources to reduce the pollution to zero?" Bogard asked about the "How clean is clean" problem. The standards should be decided by science itself, and should not be affected by politics. She complained, however, of the companies' responses, because the companies contradicted each other over responsibility for the pollution and tended not to disclose vital information.

On the other hand, citizens are not happy with Ms. Bogard's opinions about an "acceptable risk". Health standards themselves have to be based on health data alone, not cost. One part per billion of 1,1,1-trichloroethane in a liter of water is the equivalent of a million molecules. And, after all, during the first trimester of pregnancy a fetus undergoes enormously rapid cell division when the infant's heart is forming; those million molecules may spread equally rapidly.

3.3.4 Jail Sentences for Environmental Violators

Prosecutors are getting tougher on environmental violators. Recently, two corporate officers in Silicon Valley, Universal Semiconductor and Golden State Circuit, faced criminal liability under environmental laws. From the start, the Clean Water Act and the Resource Conservation and Recovery Act (RCRA) contained versions of the "knowing endangerment" rule, which provides for penalties or imprisonment (or both). However, since the federal sentencing guidelines of November 1987 were instituted, the number of criminal prosecutions has increased. In Silicon Valley, the District Attorney's environmental crimes unit has investigated approximately 100 violations of local and state hazardous waste laws a year. In 1990, complaints have resulted in five jail sentences and approximately 60 fines.

3.3.5 Water Shortage

Yosemite is a famous national park in the USA. Hetch Hetchy Reservoir in Yosemite supplies 20% of the water used in Silicon Valley. While Hetch Hetchy water contains only approximately 30 mg per liter of particles, Santa Clara's two underground wells contain 300–500 mg particles per liter. It is much more difficult and expensive for business to purify well water.

Over the past 5 years, pumping for cleanup has caused the groundwater level in South San Jose to drop by nearly 30 ft. Faced with such a drastic decline in groundwater supply, the RWQCB gave IBM and Fairchild permission to leave more contaminated water in the ground. That angered both environmentalists and Betty Roeder, who owns the Great Oaks Water Company. She has refused to supply her customers with water that contains any detectable levels of industrial chemicals. Despite the board's arguments, even Fairchild representatives said they are reluctant to slow the pumping. They fear Fairchild could face lawsuits (*San Jose Mercury News*, February 6, 1988). Because high-tech industries use so much high-quality water, they have to face a water shortage and the skyrocketing costs for water. In March 1990, IBM built a sophisticated water-treatment plant to pump water from a contaminated aquifer, to use it, and to then reinject clean water back into the ground.

3.4 What Can We Learn from Silicon Valley?

3.4.1 Waste Reduction[29]

I would first like to mention two recent developments in tackling the cleanup, and then attempt to sum up what we can learn from Silicon Valley.

1. The enforcement to tackle waste reduction. Because of the strict federal regulations on waste landfill levels, industries have to locate new disposal sites or introduce new treatment methods. In Santa Clara County, approximately 100 000 tons of hazardous waste are generated each year. If this waste was reduced at the site, the number of storage tanks and the possibility of leakage from tanks would decrease. The county is now preparing to encourage waste reduction.

2. The EPA's Technical Assistance Grant Program. Recognizing the need for citizens to be well informed of the conditions and activities at Superfund sites in their community, and the importance of informed comment from citizens, Congress established the Technical Assistance Grant Program as part of the Superfund program. The Technical Assistance Grant Program provides grants to citizens groups to obtain assistance in interpreting information related to the cleanup at Superfund sites. In February 1989, the EPA awarded Silicon Valley Toxics Coalition a Technical Grant amounting to US$50 000, one of the first in the nation. The Coalition then proceeded to hire two consultants, who commenced work to evaluate the IBM site. In April 1993, the US$100 000 award was granted. The grant will allow the Coalition to hire a technical advisor to assist it in participating in the management of the investigation and clean-up activities at the Moffett Field Naval Air Station Superfund site and the adjoining Middlefield-Ellis-Whisman Superfund site in Mountain View.

3.4.2 Lessons from Silicon Valley[30]

We have to learn many lessons from Silicon Valley.

First, pollution is vast and very serious. The cost of the cleanup is enormous. Total recovery is impossible. Groundwater pollution by storage tanks is especially dangerous. In the case of Japan, although groundwater contamination is already confirmed, thorough investigation has been delayed. It is absolutely necessary to examine not only the exhausted gas holes and drainage systems, but also underground storage tanks and surrounding wells.

Second, because high-tech pollution is, by nature, a complex chemical reaction, the focus must be on the management of chemical substances. It is desirable for industries to disclose information about the materials they use. We have to pay attention to the fact that, in Silicon Valley, people came rightly to suspect the harmfulness of 1,1,1-trichloroethane, which in Japan is regarded as comparatively safe.

Third, the concentration of high-tech industries poses potential hazards to local residents, because the storage of many sorts of chemicals and gases is highly concentrated, and these must be transported into and out of the industrial site. Safeguards against fire, earthquake, and accident are therefore indispensable. Even in the case of a planned location for a facility different from Silicon Valley, such safeguards are necessary.

Fourth, it is very important for governments to disclose their information and to allow citizens to organize, clean up, and control pollution. In Silicon Valley, the involvement of the citizens has stimulated all levels of government to take action in a relatively short time. And, although it is true that not all the problems have been solved, the involvement of citizens indicates a good model for Japanese environmental policy. In Japan, environmental surveys released by government departments generally include no names of specific factories, while hearings from residents are often perfunctory.

Fifth, I would like to propose that there is a relationship between the environmental problem and the trade friction problem. I have often heard that because Japanese companies steal technical information, American companies cannot disclose the names of the chemicals in the factories, or that competition with Japan hampers investment in safety. Japan is made an excuse for oppressing American citizens. It is high time that the citizens of both countries exchange information, especially concerning safety and problems of health.

4. The Real Semiconductor Industry

4.1 Characteristics of the Semiconductor Industry

4.1.1 What is a Semiconductor?

I would like to offer a brief description of an integrated circuit (IC) and a large-scale integrated circuit (LSI), the pivots of the high-tech industry. There are three component parts: the first is a conductor that carries electricity; the second is an insulator that does not carry electricity; and the third is a semiconductor that will carry electricity if the conditions are appropriate. A semiconductor carries electricity when it receives light or heat; this means that a semiconductor changes its character from "not carrying electricity" to "carrying electricity" depending on conditions. The convenient character of a semiconductor enables it to memorize letters and numbers and to compute quickly.

A semiconductor computes by the binary scale of only 1 and 0.1 represents the "on" of electricity and 0 represents the "off" of electricity. A semiconductor switches very fast from one to another, 10–20 million times per second. A semiconductor does not memorize letters and figures as they are, but memorizes by cutting them at right angles, each black part and white part corresponding to an electrical "on" and "off". When a special sort of semiconductor receives electricity, it activates the movements of an inner electron and emits lights at a high level of energy. This is a light-emitting diode (LED). There are many sorts of semiconductor that synthesize sounds or respond to lights.

4.1.2 "A Visit" to a Semiconductor Plant

I would like to invite you to "visit" a semiconductor plant. The semiconductor manufacturing process consists of two main steps: the first is wafer manufacturing; and the second is assembly and inspection. The wafer is fabricated by introducing impurities (dopants) to purified silicon, which may sound a paradox.

First, a silicon or gallium arsenide ingot is "grown" by chemical process (crystal growing). The semiconductor ingot is shaped and sliced into thin wafers (ingot grinding and sawing). The wafers are treated with chemicals to give them a very clean and smooth surface (wafer preparation). A thin, non-conductive layer of silicon dioxide is "grown" on the wafer (epitaxy and oxidation). The desired pattern for the electrical circuit is etched into the wafer with a set of photo masks and ultraviolet light (photo masking and etching). The exposed patterns are doped at high temperatures to result in electrical circuits in each "chip" on the wafer (diffusion). These processes are repeated.

To avoid the settling of dust, which would degrade a semiconductor, cleanliness is essential during all parts of the manufacturing process. Operatives, for example, are not allowed to wear makeup, and one Japanese semiconductor manufacturer, who merged with a Silicon Valley company, directed his workers, who came from 18 countries, to change their shoes and clothes upon arrival at the plant, and to wash their hands immediately, in order to raise the percentage of good products.

Therefore, you can easily understand that a wafer-manufacturing plant is more like a chemical laboratory than an electronics factory. In general, wafer manufacturing requires large quantities of water, electricity, and chemicals. At the same time, the assembly line requires a relatively large work force. Moreover, overnight labor for the wafer process is indispensable. The assembly and inspection processes use chemicals such as organic solvents and radioactive substances like krypton 85.

Many assembly lines belonging to American and Japanese semiconductor companies are now located in Southeast Asia. In pursuit of good water quality and cheap labor, many semiconductor facilities are located in Kyushu (Japan's most southern island). Japanese research and development facilities, however, are concentrated near Tokyo or in the USA.

4.1.3 A Comparative Study of Japan and USA Semiconductor Industries

Trade friction between semiconductor businesses is a typical economic issue dividing Japan and the USA. The USA semiconductor industry consists in the main of: (1) captive manufacturers, like IBM, AT and T, and Hewlett Packard; and (2) specialized semiconductor manufacturers, for example, 170 companies in Silicon Valley. On the other hand, the main Japanese computer makers, like NEC, Fujitsu, Hitachi, and Toshiba, themselves both manufacture semiconductors and sell them abroad. The Semiconductor Industry Association (SIA) is an organization of relatively small-scale semiconductor manufacturers in the USA.

The Japanese semiconductor industry has developed its share of business mainly in the field of the memory chip LSI, such as DRAM (dynamic random access memory).

On the other hand, the American semiconductor industry has specialized in the technologically more valuable fields of microprocessors and ASIC (application-specific IC). Today, Japanese manufactures are paying a great deal of patent royalties for DRAM and microprocessors to manufacturers in the USA, for example, to Texas Instruments. The major USA semiconductor manufacturers have developed global strategies and have located their facilities in Europe, Asia, and Japan. The market shares of the Japanese and American semiconductor industries reflect these factors: in 1984, Japan's share in the USA market, including captive USA manufacturers, was 9.6%. On the other hand, the USA's share in the Japanese market, including shipment from USA companies in Japan and other countries, was 19.1%. At the same time, because of the semiconductor treaty between Japan and the USA, the USA's share in Japan is expanding. The direct market share of the USA semiconductor business in Japan increased from 8.4% in 1986 to 20.2% in 1992. Therefore, we have to look carefully at the fact that, contrary to general belief, a strong incursion has been made into Japan.

It is more significant that the Japanese and USA semiconductor companies have developed commonly agreed strategies of technology and marketing. It is also noteworthy that a multinational company like Texas Instruments has established "twin" factories in Japan and USA, developing common methods to raise the rate of good products, to keep costs down, and to communicate information and circuit maps by communication satellite.

Because the research, development, and manufacturing of semiconductors now operate on a global scale, citizens environmental protection movements cannot cope with the situation without themselves having to develop worldwide networks to communicate their information.

4.2 Relationship Between High-Tech and Environmental Problems[31]

I have already pointed out that the relationship between high-technology and environmental problems focuses on high-technology like microelectronics and new material, while biotechnology develops on the basis of new sorts of substances: this is contrary to the saying "the message is more important than the material". These substances, even if they have little value in themselves, have long-term and combined effects on human health that are not yet sufficiently clear. At the same time, it is very hard to dispose of some of these materials. Still more, the severe competition between companies accelerates high-technology related research and development. Because information about high-technology is restricted, protective environmental action tends to be delayed.

4.2.1 The Semiconductor Production Process and Environmental Problems

When I analyze the environmental aspects of semiconductor production processes, two problems come to the fore: (1) why do manufacturers use so many sorts of chemicals and gases, including toxic gas; and (2) is it possible to make the production of semiconductors safer?

Many of the processes used in semiconductor fabrication involve chemical reactions for microprocessing and multilayering. These reactions include plasma reactions and ion implantation, where ion and plasma coexist and the free radical of high chemical activity generates itself.

Microelectronic device fabrication used to involve the use of photoresist chemicals before biologic scientists and physicians began to learn about their toxicity. The toxicity of the plastic monomers is usually much greater than that of the finished polymers. For example, vinyl chloride monomers are much more toxic than polyvinyl chloride. Little is known of the toxicological properties of the photoactive components.

The microscopic precision necessary for producing IC depends on several processes that utilize radio frequency and microwave radiation as well as X-rays. Some investigators have wondered about the relative harmfulness of radio frequency radiation of this kind. The severe competition between semiconductor companies stimulates new processes before related technology has been developed to monitor them. Hazardous materials are often used simply because they were the first to be utilized in the research laboratory.

4.3 The Material Balance of a Semiconductor Plant

How many materials that affect the environment do semiconductor factories in Japan use? Basing my analysis on *A Survey of IC Industry and the Environmental Protection* (1987) by four Japanese ministries, which includes a questionnaire to nationwide semiconductor plants (1985) and inspection (10 plants), I should like to enumerate the chemicals used in each process and describe how they are treated and disposed of.

4.3.1 Water Usage

During the summer of 1987, many semiconductor plants in the Tokyo area were forced to limit their operations because of a water shortage. Semiconductor fabrication uses a great deal of water for air-conditioning and cleaning. Many plants pump up groundwater. According to the *Survey*, one semiconductor plant uses 1.6 million gallons on average per day (minimum 0.4 million gallons, maximum 4 million gallons per day). MITI's *Statistics of Industry, Lands and Water* (1991) reveals that one IC plant uses, on average, 2.7 million gallons per day. Maximum amounts are 10-fold greater than minimum amounts. The amount of water used by each plant depends on the conditions of the area in which it is located.

Some plants, like Texas Instruments, Miho (Ibaragi Prefecture) and Japan Foundry, Tateyama (Chiba Prefecture), implement a "closed system" of drainage. In spite of higher investment and running costs than usual, the closed system has some advantages: (1) more strict regulation of the water quality is required; (2) the use of too much water is difficult; (3) more hazardous materials are used in the process; and (4) the closed system offers a good impression of the plant.

4.4 Organic Solvent

According to the *Survey*, trichloroethylene—the organic solvent most extensively used in the wafer-making process—amounts to 8.4 million pounds (51 plants) a year. Seventy percent of the trichloroethylene is used for washing away photoresist; 30% is used for cleaning of chips and instruments; 20% of the input becomes exhaust gas,

78% of which is discharged without treatment; and 80% of the input becomes liquid, two-thirds of which is sold for recycling, and one-third of which is disposed of by "trader on commission". On the other hand, the amount of trichloroethane used is very small (410000 pounds). Forty-eight percent of the input becomes exhaust gas, 80% of which is discharged without treatment; and 50% of the input becomes liquid, 30% of which is sold for recycling use, and 70% of which is disposed by trader on commission.

Therefore, considerable amounts of organic solvent are used during semiconductor fabrication. Inspections (10 plants) detected a relatively high density of organic solvents near exhausts without any treatment apparatus.

An investigation (1988) by the Environment Agency of Japan reported that air pollution by organic solvents is at a critical stage. In particular, tetrachloroethylene and carbon tetrachloride were detected at 37 points nationwide. The maximum detected levels of trichloroethylene, tetrachloroethylene, and carbon tetrachloride are above the regulation level set by the WHO. The Air Pollution Control Law of Japan, however, does not designate organic solvent as a toxic substance. Several options are open to semiconductor manufacturers to reduce these emissions and these can be divided into three major categories: (1) the addition of control equipment; (2) the modification of material reformulation process; and (3) the improvement of manufacturing procedures. Emissions from positive photoresist operations are one-tenth of those from negative photoresist operations.

4.4.1 Treatment of Exhaust Gas and Waste Disposal

As to the treatment of the exhaust gas itself, trichloroethylene is treated with a water scrubber (aeration apparatus); however, this process is not very efficient. Activated carbon removes up to 90%–95% of the trichloroethylene. Solid and liquid wastes of organic solvents and toxic gases are mainly disposed of by trader on commission. Inspections at the ion implantation process (10 plants nationwide) detected trichloroethylene in the waste oil and arsenic in the vacuum pump oil and on clothes.

Near Tokyo, some cases of illegally dumped organic solvent waste have been found. Leakage from waste solvent transportation trucks is another problem. In February 1985, in Ibaragi Prefecture, a 212-gallon leakage of waste solvent from a truck polluted a rice field, and halted its operation.

4.5 Problem of Organic Solvent

Organic solvents are widely used not only by the semiconductor industry, but also by laundries, for degreasing metal, and as painting solvents. These organic solvents include trichloroethylene, tetrachloroethylene, trichloroethane, and cellosolve (EGE). In the past, the primary concern about solvents focused on their acute irritating effects on mucous membranes. More recently, attention has shifted onto their chronic, neurologic, and neuropsychologic effects, as well as their carcinogenicity and their harmful effects on human reproductive processes. Because an air-conditioner in a "clean room" gets rid of dust but not evaporated chemicals, the air-conditioner can circulate these chemicals within the plant and, thus, expose the workers to recirculated solvent and chemicals.

4.5.1 Trichloroethylene

As the *Survey of IC Industry and the Environmental Protection* by four Japanese ministries (1987) shows, the percentage of trichloroethylene used in Japan is still high. As to its effects on the central nervous system, depression, dizziness, lack of coordination, loss of muscle control, drowsiness, fatigue, peripheral neuropathy, tremors, giddiness, anxiety, nausea, vomiting, and behavioral changes have all occurred in animals exposed to concentrations as low as 9 ppm.

In Woburn, Massachusetts, as I have already noted, a group from the Harvard School of Public Health found a positive statistical association between access to water contaminated with trichloroethylene and the incidence rates of childhood leukemia. For people exposed to the solvent, a new theory stresses the need for attention to the role played by taking a shower and washing one's hands in addition to the monitoring of drinking water.

4.5.2 Tetrachloroethylene

When you walk past a laundry shop, you can smell a special odor, that of tetrachloroethylene, used in dry cleaning. The semiconductor industry also uses tetrachloroethylene. A recent epidemiologic study of dry cleaning workers (1711 persons) showed an excess of deaths from "other forms of heart disease" and "other disease of the liver" among dry cleaning workers with a history of exposure to organic solvent, regardless of personal habits. Because dry cleaning in Japan also uses petroleum solvents and 1,1,1-trichloroethane, we cannot attribute these diseases to tetrachloroethylene only. Nevertheless, we have to pay attention to the results of this study.

4.5.3 1,1,1-Trichloroethane (Methylchloroform)

On our desks, we keep bottles of white liquid that we use for making corrections; this liquid includes 100% pure 1,1,1-trichloroethane. Because 1,1,1-trichloroethane is regarded as less harmful than trichloroethylene, Japanese industries have replaced trichloroethylene with 1,1,1-trichloroethane. However, as I remarked in section 1, 1,1,1-trichloroethane is suspected of causing birth defects. A group from the New Jersey Medical School investigated the relationship between 1,1,1-trichloroethane exposure and the appearance of cardiovascular abnormalities in rats. Furthermore, 1,1,1-trichloroethane helps to deplete the ozone layer, although its level is not so high as that of CFCs (freon gas). The reported release of 1,1,1-trichloroethane into the Californian environment has increased 10% in 3 years. This solvent is used widely in the aerospace, auto, metal products, electronics, and chemical industries.

4.5.4 Glycol Ether (EGE)

The semiconductor industry also uses glycol ether, cellosolve, as a photoresist. The State of California recently recommended that methoxyethanol and ethoxyethanol should be regarded as potential toxins on male and female reproductive organs after animal studies demonstrated sperm toxicity, embryotoxicity, and teratogenesis. These effects occurred at levels close to the current permissible exposure levels (PELs). In Japan, too, the PELs of EGE have been lowered from 200 to 5 ppm. The Japanese Association of Occupational Safety, Hygiene Investigation Center, has carried out animal studies on EGE; the results make clear the need for much stricter regulations than are in force at present.

4.6 Problems of CFCs

At Berkeley and Palo Alto, California, containers for takeout meals made with CFCs are banned. CFCs, or chlorofluorocarbons, which consist of chlorine, fluorine, and carbons atoms, are non-toxic, inert, and degreasing. Today, CFCs are used in refrigerators and air-conditioners, as blowing agents to create foam in containers and cushions, as solvents to clean computer chips, and as propellants in some aerosol cans. High-tech companies, especially, use a rapidly increasing amount of CFC-113. American electronics manufacturers consumed approximately 55.5 million pounds of CFC-113 in 1984, 37% of all CFC-113 produced that year. IBM was the biggest electronics user of CFCs in the world. For example, IBM's San Jose plant purchased more than 2.6 million pounds of CFC in 1987. When exposed to sunlight, the chlorine in CFCs can destroy ozone molecules in the stratosphere. Ozone molecules absorb most of the ultraviolet radiation that comes from the sun, while ultraviolet causes sunburn, some skin cancers, and effects changes to the climate and within the earth's ecosystem.

Furthermore, each CFC molecule is 20000-fold as efficient at trapping heat as one molecule of CO_2. CFCs therefore augment the Greenhouse Effect. CFC molecules have a long life span, approximately 100 years. Their concentration in the air in Japan has risen 2.5-fold in 7 years. In Japan, approximately half the CFCs utilized are used for cleaning (mainly high-tech industry).

In September 1987, the Montreal Protocol proposed a 35% net reduction of CFC production worldwide by 1999. In 1992, representatives of 86 countries declared their intention to phase out their production and use of CFCs by 1995; it was also decided to phase out 1,1,1-trichloroethane by 1995.

4.7 Toxic Gas[32]

4.7.1 Dangerous Toxic Gas

I mentioned in section 2 that, in March 1988, three people were killed in New Jersey by silane gas. In Japan, as well, on October 2, 1991, two people were killed by silane gas required for the chemical vapor deposition (CVD) apparatus at Osaka University. The danger and toxicity of gases used in semiconductor fabrication come from their flammability, corrosiveness, explosiveness, suffocating nature, and direct toxicity.

As to flammability, for example, silane is a pyrophoric gas, which even 100% halone gas cannot extinguish. In Japan, in November 1984, 22 pounds germane (GeH_4) exploded at Nippon Sanso, Kawasaki (Kanagawa Prefecture) as the result of a decompositive reaction, while in December 1989, an explosion, probably caused by monosilane, killed one person and injured three persons at a research and development facility belonging to Hitachi, Musashi (Tokyo).

In Japan, 17 cases of accidents (1976–1988) caused by toxic gas in the semiconductor industry have already been reported officially by the Ministry of Labor. Approximately half these cases involved death by suffocation after the inhalation of inert gases. Inert gases, like nitrogen and argon, are used as balance gases. Suffocation is partly due to the mistaken belief that a "clean room" circulates air and that "nitrogen"

TABLE 4.5. Toxic gases classified according to their dangerousness

Flammability	Germane, arsine, phosphine, silane, diborane, dichlorosilane
Explosiveness	Hydrogen, ethylene, acetylene, propane, hydrogen sulfide, carbon monoxide
Suffocation	Argon, nitrogen, carbon dioxide
Toxicity	Arsine, diborane, phosphine, silane, boron trifluoride, ammonia, carbon monoxide, hydrogen sulfide, chlorine hydrochloride, boron trichloride, phosphorus trichloride, silicon tetrachloride
Corrosiveness	Ammonia, hydrochloride, chlorine, boron trichloride, phosphorus trichloride, phosphorus pentachloride, silicon, tetrachloride
Odor	Dinitrogen oxide, hydrogen sulfide, chlorine, ammonia hydrochloride, boron trifluoride

Source: Science Forum (1984) Semiconductor gas safety data book. Tokyo

means safety. Some places with a deficiency of oxygen have been found only after semiconductor plants started operating. As to toxicity, only one respiration of certain substances will cause deadly damage. Hydrides like arsine, stibine, and germane cause severe hemolytic anemia with a peripheral blood smear that may show anisocytosis, red cell fragments, and ghost cells. It is reported that the proliferative response of human peripheral lymphocytes may be a useful indicator in the evaluation of the toxicity of arsenic, for instance, while workers at wafer processes have a relative enhancement of their proliferative response (Table 4.5).

How can we handle gases safely? Mr. Hikaru Harada (Nippon Sanso) proposes "four principles of safety" for handling gases: (1) exact knowledge of the nature of the gas; (2) the use of safety apparatus; (3) the accurate operation of machinery; and (4) safety training.

1. "Exact knowledge of the nature of the gas" means that we should not use any sort of gas without an exact knowledge of the nature of each gas. It is noteworthy that harmfulness to human health depends on whether the element's form is either a simple substance, an oxide, or a hydride.

2. "Use of safety apparatus": it is necessary to reduce the number of connecting parts and, thus, the possibility of leakage, and to select fireproof and corrosion-proof material. For example, in the case of a fire at Miyazaki-Oki (Miyazaki Prefecture) in 1982, pipelines made of vinyl chloride connected to a CVD apparatus were the cause of the fire. Some Japanese chemical manufacturers who have recently commenced semiconductor gas fabrication, have tried to break into the market without installing sufficient pieces of safety apparatus, in an effort to keep down prices and reduce the safety control costs. In the USA, major chemical manufacturers are also beginning to operate in this way. A growing number of semiconductor companies use 100% phosphine in the doping process. This high concentration of phosphine is not necessary, but is merely a convenience to avoid changing cylinders and, thus, to keep down costs, because cylinders of lower concentrations of phosphine would be more expensive to run.

According to the *Report of Specialty Gas Usage* issued by Kanagawa Prefecture, Japan (61 plants; 1985), approximately half the plants have no special power sources for gas use, and over 80% of plants keep cylinders on site. Small-scale factories handling only one to three sorts of gas have particular problems.

3. "Accurate operation of machinery": *The Survey on Toxicity and Use of Gases used by Semiconductor Fabrication* (76 plants; 1987) issued by the Ministry of Labor claims that 70% of the near-miss accidents are due to a malfunction, whereas 30% are due to insufficient inspection of the machines. Accidents related to the changing and transporting of metal cylinders offer an example: (1) mistaking safety valves for opening valves; (2) toxic gas leakage at the time of cylinder washing; (3) connecting cylinders that have different container concentrations; and (4) the dropping of cylinders.

4. "Safety training" is, therefore, very important. However, small-scale businesses and captive manufacturers (e.g., watch makers) do not have enough means to undertake such training.

In all cases, the semiconductor industry should seek to replace hazardous processes and materials with safer ones. In fact, some manufacturers are trying to use safer processes. For example, silicon tetrachloride is now used instead of silane, organic arsenic is used instead of arsine, and phosphorus chloride is used instead of phosphine.

4.7.2 Exposure to Dopant

Investigators from the National Institute of Occupational Safety and Health (US NIOSH) have recently shown that low levels of dopant compounds are released from silicon wafers for several hours following ion implantation. This may account for the "odor problems" that workers often complain of. Thus, for example, Suspect Inhalation from Doctor First Report of Illness for Santa Clara and San Mateo Counties, October 1978–March 1980[33] shows that the dopants are in the first rank as suspect causative agents. A "clean room" is, in general, well ventilated, but ventilation in certain localities within a room is not adequate.

4.7.3 Radio Frequency and Microwave Radiation

Today, many families use radio frequency ranges (cooking stoves). However, besides producing heat, radio frequency has many biologic effects, such as mutagenesis, carcinogenesis, reproductive disorders, growth and development of teratogenicity, ocular disturbances, neurologic and behavioral disorders, and cardiovascular problems. Recently, the plasma process has become more common for removing coating (photoresist etc.) from wafers. Researchers in the USA surveyed four microelectronics manufacturing facilities and found up to 61% of the plasma etchers produced external radio frequency (13.56 MHz) exposures in excess of recommended standards.

4.7.4 Potential Disaster

Although high-tech plants appear to be "lean", "smart", and "fireproof", semiconductor plants nevertheless use many sorts of toxic chemicals at high temperatures and high voltages. If there is a gas leak, a power failure, or a machine stoppage, these plants risk a big fire or a major disaster. If a "clean room" catches fire, it cannot receive pump-in water from outside if there are no windows. Air ventilation may circulate toxic gas. The concentration of apparatus that uses toxic chemicals hinders efforts to extinguish fires. An unbelievable accident occurred one day when it was found impossible to get water to the fire. This happened at the Miyazaki-Oki LSI plant (Miyazaki Prefecture) in October 1982. The origin of the fire, which killed one person, seemed to be a CVD apparatus.

4.8 Gallium Arsenide[34]

These days, we can watch a large-scale video at a baseball stadium. This type of video uses LEDs made of gallium arsenide: and gallium arsenide has emerged as an alternative semiconductor material to silicon. The advantages of gallium arsenide over silicon are: (1) high speed; (2) low power; and (3) radiation and temperature resistance. Furthermore, a combination of gallium arsenide and aluminum gallium arsenide, the high electron mobility transistor (HEMT), has been developed as a high-speed chip for a broadcasting satellite tuner. In Japan, some wafer makers, like Sumitomo Denko, and Mitsubishi Monsanto, and a few other device makers, like Fujitsu and SONY, are now manufacturing gallium arsenide wafers.

A recent case study shows that extremely high levels of arsenic were found in the cut-off room of gallium arsenide, while high levels of arsenic were also found during a liquid-encapsulated Czochralski (LEC) cleaning operation. Other studies indicate that the dissolution of a gallium arsenide compound may occur in vitro and be associated with significant blood arsenic concentrations in vivo following intratracheal and oral administration. Respirable gallium arsenide dust in ingot and wafer fabrication may be considered a source of arsenic exposure.[35] Therefore, we must pay just as much attention to the components and products of gallium arsenide as to other materials.

4.9 Economic Aspects of High-Tech Pollution

Needless to say, the competitive development of high-technology is a capitalist business activity, and, therefore, the environmental problems caused by high-tech development have an economic background. In my book *The Economics of Environment and Technology*,[36] I have analyzed the relationship between business activities and environmental problems. My main conclusion were these:

1. Because of the pursuit of profit and "cost-down" by businesses, savings on environmental protection apparatus causes environmental disruption;
2. Severe competition for new brands and chemical developments within a limited time by big companies forces them to omit the installation of safety checks of new products;
3. Therefore, it becomes a matter of urgency to devise systems to control and monitor artificially created synthetic chemicals;
4. Making use of "free natural reserves", such as groundwater, leads to the ruthless exploitation of natural resources; and
5. Business companies, under social pressure to attend to environmental protection, try to recover profits by pursuing "cost-down" by recycling and reduction of waste.

I would like to analyze high-tech pollution in relation to these factors.

First, formerly, environmental questions used to deal with drains, exhaust gases, industrial waste, and garbage. Recent focus on high-tech pollution shows, however, that leakage from underground solvent tanks has caused groundwater pollution. This phenomenon will not have happened by accident. Why are there so many leakages from plants? The problem seems to be that many underground tanks are made of fiberglass, which is susceptible to chemical corrosion and will crack under uneven ground pressure. As I mentioned in section 3, in Silicon Valley, today, double containment and strict monitoring are implemented. It is therefore clear that cost savings on safety management, such as pressure testing and waste treatment, caused the leakages from underground tanks.

Second, new production methods, storage methods, and the use of many sorts of chemicals can cause environmental problems. New high-tech products are developed one after another, and their production processes are constantly being changed. Chemicals used for these processes are also developed under conditions of trade secrecy. These conditions necessitate a safety check system for new chemical products, and the gathering and disclosure of data at the level of government and companies. Furthermore, the use and storage of chemicals remains a problem. As I have already said, some chemical manufacturers who have recently entered into the business of semiconductor fabrication try to break into the market without equipping their factories with sufficient safety apparatus in an effort to keep down prices; this arises from a desire to reduce safety control costs in that industry. Small-scale business and captive manufacturers do not have enough financial means to carry out safety training procedures.

Third, high-tech pollution abuses groundwater by pumping up vast amounts of water and contaminating the groundwater that remains in the ground. Because underground microbes dissolve and absorb pollutants, groundwater has formerly been regarded as safe, and the protection of groundwater was regarded as a secondary subject. However, the pollution of groundwater draws attention to the importance of groundwater as a drinking water resource. The protection of the quality of groundwater, as well as its quantity, becomes urgent. As I have already observed, one Japanese semiconductor plant uses, on average, 2.7 million gallons of water per day. This is because the industrial costs for groundwater are lower than those for surface water. However, because of the location of plants and the agreements for environmental protection made between plants and cities, some plants do not use groundwater (e.g., Iwate-Toshiba Electronics, Iwate Prefecture) and other plants adopt closed systems of drainage (e.g., Japan Foundry, Tateyama, Chiba Prefecture). Today, some plants are trying to recover organic solvents and CFCs as well as drainage. Under the pressure of regulations, citizens movements, and local agreements, semiconductor plants are trying to save materials and water with closed systems.

Fourth, problems are caused by those who try to "shuffle off" the duty of waste disposal. These problems are: (1) soil and groundwater contamination created by drum recyclers; (2) groundwater pollution caused by leakage from the sewage drains of a special "recycle" trader of organic solvent; and (3) pollution of a landfill site by a "disposal" trader. According to data provided by the Japanese Ministry of International Trade and Industry, only 1.45% of the waste (approximately 7 billion pounds) produced in IC factories was reused in 1988. While high-tech companies may keep themselves clean, this could be because they transfer their duty to treat and dispose of waste to others, and this may, in turn, cause pollution beyond the boundaries of the high-tech industries. This practice is the result of trying to avoid paying for the cost of environmental protection. We also have to monitor and regulate this sort of practice.

5. High-Tech Pollution in Japan

5.1 Groundwater Pollution by Toshiba Components, Kimitsu

In 1984, a child whose parents lived close to groundwater was born with serious defects. During the summer of 1987, pond carp began to swell and die. These incidents were recorded in Kimitsu, Chiba Prefecture, where groundwater pollution by a semiconductor plant was found during the spring of 1987, yet was not made public until September 1988. Forty-three wells were investigated; 10 wells were found to be contaminated by trichloroethylene in quantities over the regulation level set by the WHO. These 10 wells included a well for drinking water, a municipal well (well #3) and the well for the municipal swimming pool.

The highest detected level of trichloroethylene was 10 000 ppb. (October 1988), which was 330-fold the regulation level (30 ppb.). The source of the pollution was Toshiba Components, Kimitsu, which is located beside the upper reaches of the groundwater stream. It is noteworthy that: (1) the municipal well is contaminated by trichloroethylene; (2) the source of the pollution is a branch of the high-tech industry; and (3) the fact of groundwater pollution was not disclosed for a year and half.

5.1.1 The Japanese Type of High-Tech Pollution

We already know from our experience of high-tech pollution in Silicon Valley that the "high-tech industry is not always clean". In Japan, however, despite the continuance of groundwater contamination, the relationship between groundwater pollution and high-tech industry has not, except in a few instances, become public knowledge. The groundwater pollution at Kimitsu is a typical case of high-tech pollution; it is also an exemplary case of Japanese high-tech pollution: both the company and the local government covered up the outbreak of pollution. The experiences of severe environmental disruption in Japan during the 1960s should have brought to an end this type of covering up of pollution.

5.1.2 Pollution Source: Toshiba Components

The pollution source, Toshiba Components, Kimitsu (500 employees), is a manufacturer of commutating semiconductors, and uses trichloroethylene for cleaning. Toshiba Components, Kimitsu, shares 25% of the world market for commutating semiconductors for automobiles. According to an investigation carried out by Kimitsu City, Toshiba used as much as 4.20 million pounds of trichloroethylene from 1972 to 1988, half of which was not recovered. The manager of general affairs at Toshiba Components has said, "We have covered the tanks of trichloroethylene with concrete walls, and moved the pipeline above ground" (*Asahi Shimbun*, Chiba Edition, September 9, 1988). Although there were no underground tanks, leakage nonetheless occurred at

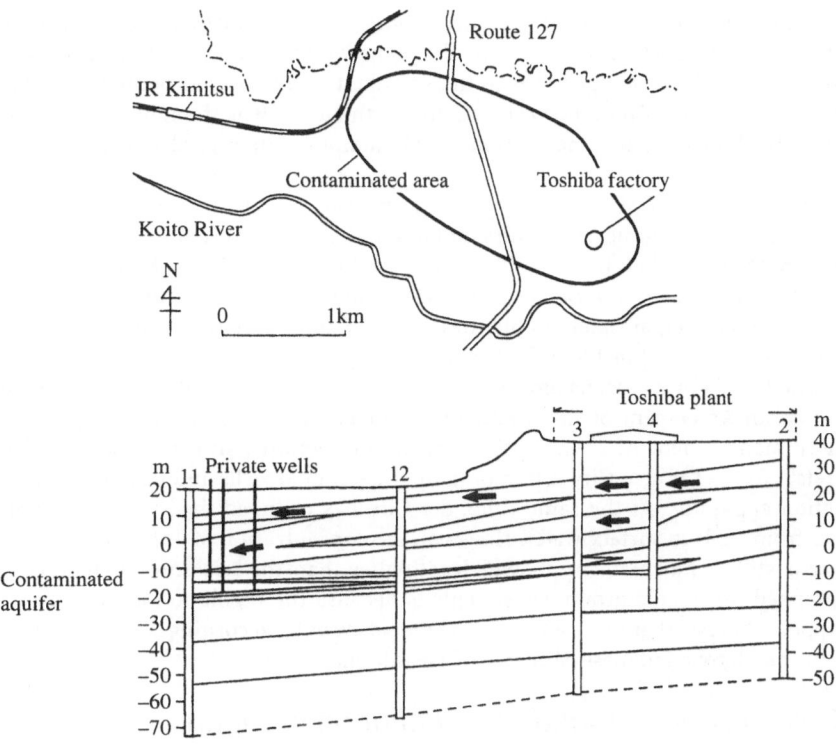

Fig. 4.2. Groundwater pollution at Kimitsu City

the time of changing and transporting of waste trichloroethylene. Trichloroethylene has been detected even as deep as 180 ft beneath the plant site.

5.1.3 City's Investigation Specified the Source of the Pollution

In January 1989, a Kimitsu City investigation, cooperating with the Water Quality Protection Institute of Chiba Prefecture, with instruments for detecting gas leakage, showed that Toshiba had polluted 70 000 cubic feet of soil to an average depth of 60–100 ft beneath the plant site.

This investigation is the most detailed survey of groundwater pollution in Japan. The investigation suggested that the causes of the pollution were: (1) the dumping of waste trichloroethylene at the disposal site; and (2) the leakage of trichloroethylene during the off-loading of materials at the site (Fig. 4.2).

5.1.4 Health Injury

Citizens in the neighborhood had complained of various health problems and required the City Hall to carry out a health survey. For instance, the child with a birth defect was hospitalized and other people have experienced miscarriages, and heart disease. In 1988, three aged persons died from unknown causes. A worker who dealt with waste trichloroethylene at the Toshiba plant was blinded when the liquid squirted into his eye. Toshiba changed over to the so-called "safer" trichloroethane in November 1988. In Silicon Valley, however, trichloroethane itself is thought to have caused various birth defects. Toshiba's actions ought simply to be regarded as a "one round delay".

A Health Inquiry Task Force set up by Kimitsu City has already carried out a simple survey of citizens, but not yet a full epidemiological examination.

It is noteworthy that, in 1984, the same Toshiba, Taishi (Hyogo Prefecture), semiconductor plant caused serious groundwater pollution. At this time, the pollution reached the municipal water supply, and levels of contamination over the regulation standards were found in 128 wells (see below).

Pushed to act by a citizens movement, in September 1989 Kimitsu City and Toshiba drew up an Agreement of Environmental Protection, which includes items such as "safety management of chemicals", "citizens inspection", and disclosure of "trade secrets". Today, Kimitsu City is pumping up the polluted wells and treating them with aeration apparatus. At the same time, the City has changed the source of drinking water from wells to surface water. However, the concentration of THMs in the surface water is still as much as 36 ppb, which is higher than the trichloroethylene level of 5 ppb fixed for treated groundwater. This is because the regulation level of THMs is 100 ppb, whereas that of trichloroethylene is 30 ppb, according to the standards established by the Japanese Ministry of Health and Welfare.

5.2 Nationwide Groundwater Pollution in Japan[37,38]

Early in 1983, a first-time survey carried out by the Environment Agency showed that groundwater throughout Japan is polluted. According to the Land Agency, 23% of all drinking water depends on groundwater. Even the fountain called *meisui* near the foot of Mt. Fuji, famous for its good quality, is contaminated by organic solvent. Data published in 1988 reveals that 22 of Japan's 48 prefectures have detected pollution of the groundwater. Furthermore, the survey of 1989 shows that 112 of 359 wells (six prefectures) include samples of *cis*-dichloroethylene, which is more harmful than trichloroethylene.

Because metal degreasing and dry cleaning operations, as well as the semiconductor industry, use organic solvent, it is not clear that the high-tech industry is the sole cause of groundwater pollution. I have therefore had to analyze the survey of cities that house semiconductor plants (wafer process), and although I find that wells in Higashine (Yamagata Prefecture), Aizu-Wakamatsu (Fukushima Prefecture), Matsudo (Chiba Prefecture), Kyoto, and Yohkaichi (Shiga Prefecture) are polluted by trichloroethylene, the relationship between the presence of a semiconductor plant and the pollution of wells remains unclear.

5.3 Toshiba, Taishi Plant—The First Example of High-Tech Pollution

5.3.1 Detection of Pollution

Ever since Shotoku Taishi (Prince Shotoku) ruled the area centuries ago, this city Taishi has been known by that name. Taishi, located to the west of Osaka, is known

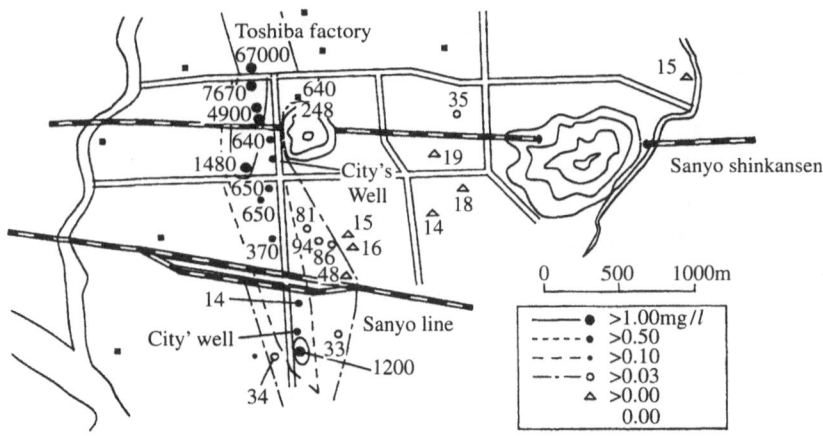

FIG. 4.3. Groundwater pollution at Taishi City (trichloroethylene)

for its long history and temperate weather. The first case of high-tech pollution caused by a semiconductor plant in Japan occurred at Taishi. The source of the pollution was Toshiba, Taishi (1200 employees). Opened in 1959, the company produces semiconductors and cathode ray tubes (CRT) for TV. Today, Toshiba, Taishi is a center for discrete-type semiconductor manufacturing. Because Taishi has plentiful groundwater, people have always utilized the wells. In 1983, an investigation for THMs detected groundwater pollution. The municipal well for drinking water registered 407 ppb trichloroethylene; 128 of 427 wells showed readings over the regulation level. At Woburn, Massachusetts, health injuries have been reported with levels of 267 ppb trichloroethylene. Later, investigations by Taishi City and Hyogo Prefecture showed that the source of the pollution was located near building #407 of the Taishi plant. From 1970, Toshiba, Taishi, had been using trichloroethylene for the cleaning of semiconductors. From 1981 to 1983, the plant also used 264 000–660 000 gallons groundwater per day (Fig. 4.3).

5.3.2 The Cause of the Pollution

What caused the pollution near building #407? A detailed analysis carried out by Hyogo Prefecture reported that "...building #407 was located near a tank of trichloroethylene" and that "...soil contamination was due to problems of storage and usage of the tank". As for the removal of the contaminated soil, the investigator reported that: "Since the groundwater spring lies 23 feet beneath the surface, it is difficult to dig out more contaminated soil". The report went on to say, "Hyogo Prefecture had previously directed Toshiba not to lay the facilities under the ground", and "since the tank and pipeline have already been removed, it is difficult to investigate more accurately the cause of pollution".[39] This report leads one to suppose that the tank of trichloroethylene had, in fact, been laid under the ground and that leakage from the tank and pipe caused the pollution. It is also worth noting that the removal of the contaminated soil was stopped at a depth of 23 ft. When we consider that at the Fairchild site, San Jose, the slurry wall goes down to a depth of 130 ft, Toshiba's action

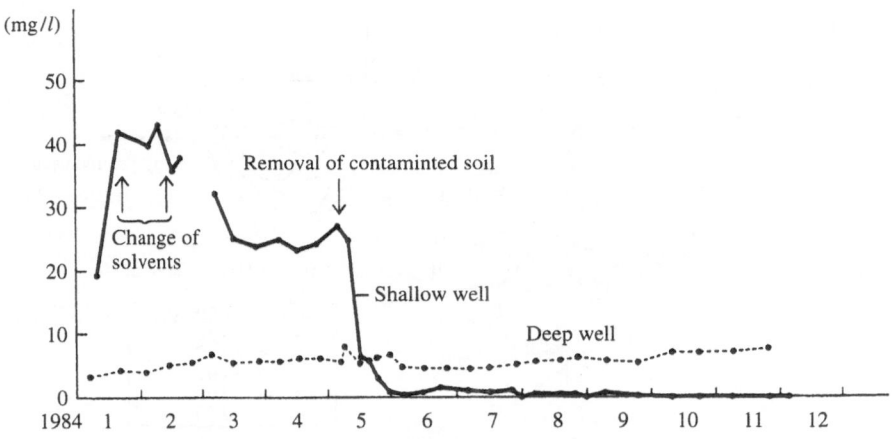

FIG. 4.4. Fluctuation of contaminated groundwater at Taishi City

is simply not thorough enough. Although after the removal of the contaminated soil in May 1984 the pollution level in the wells decreased, the contaminated soil that was not removed would still be able to pollute wells located downstream of the ground-water. The contamination levels in deep wells (to a depth of 130–200 ft) are therefore gradually rising (see Fig. 4.4) In 1988, the highest pollution levels in local wells measured 500–600 ppb. Without the removal of the contaminated soil to depths below 23 ft, groundwater pollution is bound to continue.

As for leakages from underground storage tanks, Canon, Kanuma (Tochigi Prefecture), which manufactures lenses for copying machines and cameras, recently polluted waste water and soil with tetrachloroethylene (2.5 ppm in the soil). The leakage seems to have occurred from an underground storage tank used to contain tetrachloroethylene and the degradation of activated carbon used in the waste water treatment.

5.3.3 Action to Halt Pollution

Today, at Taishi, after aerating of the groundwater and treating it with activated carbon, drinking water again comes from the municipal wells. Private wells have been taken over as public water services. At the same time, Taishi City Council has discussed the safety of the water after treatment and air pollution caused by aeration. Although Toshiba changed from trichloroethylene to trichloroethane in February 1984 and attached an absorbing apparatus with activated carbon to each cleaning facility, trichloroethylene has nonetheless been detected in the surrounding wells (maximum 640 ppb.), whereas trichloroethane is being detected at even higher levels than before.

5.3.4 Problems to Be Solved

The Agreement for Environmental Protection drawn up between Taishi City and Toshiba does not include special items to cover the use of groundwater and chemicals. Although Toshiba is required to report to the City the name of the chemicals it

uses, citizens are not permitted to know what these chemicals are. Because, in July 1987, a fire broke out at a new building on the site, it has become necessary to treat the semiconductor facility as a chemical plant. Although liver complaints have been reported among local citizens, no health survey has been carried out.

Toshiba does not admit formal responsibility for the pollution; but the company pays the cost of water facilities as a "donation" (sic).

Although Hyogo Prefecture, who investigated the pollution, admitted leakage from the underground tanks as a matter of fact, it nevertheless recorded that the cause was "unknown"; it also failed to carry out a more detailed investigation of the contaminated soil or a health survey of the citizens. The owners of the private wells, uneasy and apprehensive, are now using groundwater for bathing and drinking. If the situation is left as it is, pollution from "unknown" causes is likely to continue for a long time and the local people will become "guinea pigs". Toshiba and Hyogo Prefecture will bear a heavy responsibility for any disasters that may happen.

5.4 Groundwater Pollution in Kumamoto—A Critical Situation

5.4.1 The State of the Pollution[40]

Although many semiconductor plants like NEC and Mitsubishi are now located in Kumamoto, famous for the high quality of its groundwater, the Prefecture is beginning to suffer from serious groundwater pollution, and has had to modify its policy to develop a "Technopolis" (a high-tech industry area). Kumamoto City itself, which depends on its groundwater, has found that a great deal of its groundwater is now polluted by organic solvents. In 1982, a survey of the Environment Agency detected groundwater pollution in Kumamoto for the first time. In 1987, Kumamoto City detected contamination in 47 wells at levels that were over the regulation level. Pollution has already reached a depth of 300 ft. Even the source of the municipal well has been polluted by organic solvent, although the level of pollution is still below the regulation limit. Although the City Hall has said that most of the pollutants come from dry cleaning plants, the pathway of the pollution has not been thoroughly mapped.

The groundwater basin at the foot of Mt. Aso, Kumamoto Prefecture, is also polluted by organic solvent. According to a prefectural survey, the electronics industry uses the largest quantity of organic solvents (24 plants; 1 750 000 pounds in 1986). In 1986, the recovery rate of solvents by the electronics industry was 55% (Table 4.6).

TABLE 4.6. Groundwater pollution at the groundwater basin of Kumamoto

	Well surveys		Wells recording excess
Wells at plants	Tetrachloroethylene	40	5
	Trichloroethylene	40	1
Wells near plant	Tetrachloroethylene	89	5
Drainage of plant	Tetrachloroethylene	44	5

Time of survey: June–September 1987
Source: Kumamoto Prefecture

NEC and Mitsubishi, who use a wafer in their semiconductor production process, have no underground tanks and have already stopped using organic solvents. Nevertheless, problems remain at every facility.

5.4.2 NEC, Kyushu

NEC, Kyushu (Kumamoto City; 3000 employees), is one of the biggest semiconductor plants in the world. NEC's strategy is to house ". . . the main plants for research and development near Tokyo, while local plants are designed for mass production facilities".

NEC, Kyushu, is located in the southwest part of Kumamoto City. It was opened in 1970, has many related companies, and transports 60% of its products by air. NEC,

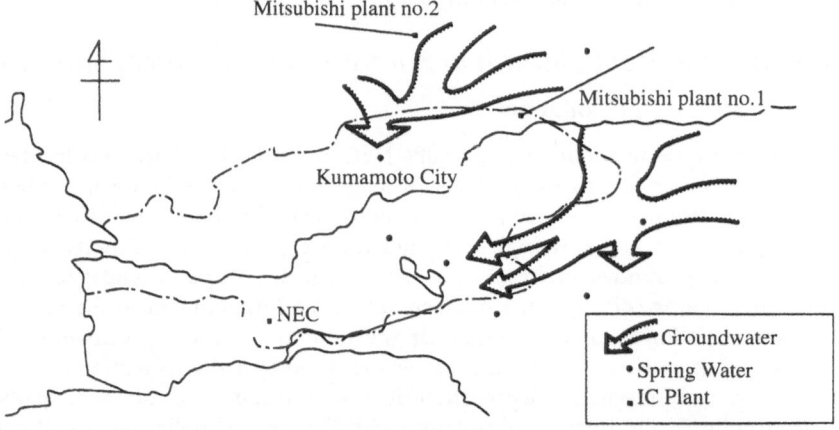

FIG. 4.5. Groundwater basin at Kumamoto City. IC, integrated circuit

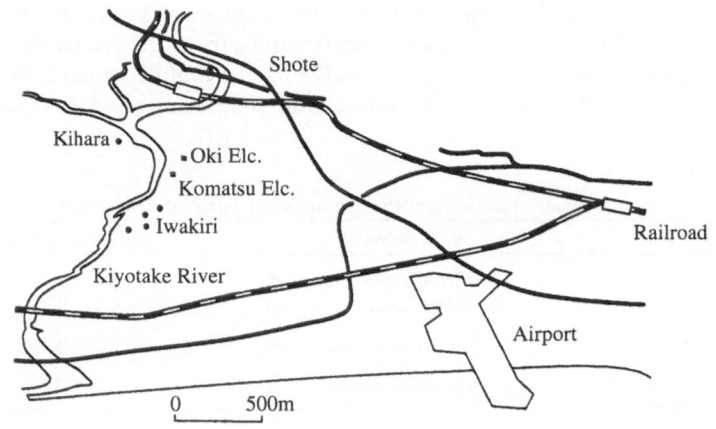

FIG. 4.6. Semiconductor plant and filtration plant at Miyazaki City. Elc., Electronics

Kyushu, once utilized organic solvents (approximately 66 000 pounds per month). The Agreement for Environmental Protection between Kumamoto City and NEC includes no regulations on underground pumping (790 million gallons per year) or on the use of toxic chemicals. The gas storage facility of Toyoko Chemicals, located to the north of NEC, Kyushu, is next door to a kindergarten and clinic.

5.4.3 Mitsubishi, Kumamoto

Mitsubishi, Kumamoto (1000 employees), is located in Nishi-Gohshi City, north of Kumamoto City, and is sited directly on the groundwater basin. According to the Regulation Section of the Environment Division, Kumamoto Prefecture, Mitsubishi has transferred its cleaning process to other companies, whose actions to ensure environmental protection are not known. The Agreement for Environmental Protection between Mitsubishi and Nishi Gohshi City contains no items relating to groundwater protection and the safe use of chemicals.

An investigation carried out by Professor Emeritus Masamoto Shimizu (December 1986) detected 2.8 and 2.6 ppb trichloroethylene near drainage outlets at NEC and Mitsubishi, respectively. It is therefore necessary to carry out thorough investigations into the pollution of groundwater in Kumamoto, and into the semiconductor facilities, especially into their use of toxic chemicals and their management of any subcontracting plants (Fig. 4.5).

5.5 Miyazaki City Municipal Water Source is Located near the Drainage Outlets of Two Semiconductor Plants

The fire at Miyazaki-Oki (1750 employees) in 1982 revealed that Miyazaki Prefecture (in Kyushu) had a semiconductor plant. Miyazaki City's municipal water sources are located near the drainage outlets of two semiconductor plants. The filtration plants at Shimokitakata and Tomiyoshi are located downstream of Kyushu-Fujitsu along the Ohyodo River, while the filtration plant at Iwakiri is located 1700 ft downstream of Miyazaki-Oki and Kyushu-Komatsu (wafer manufacturer) along the Kiyotake River.

5.5.1 The Degradation of Source Water

The Kiyotake River has two filtration plants upstream, and one, the Iwakiri filtration plant, downstream. These three plants use shallow wells (a depth of 20 ft) and river water under flows. Investigations carried out by the Japan Scientists Association, Miyazaki Branch, show that the presence of nitrate, chlorine ion, calcium, and magnesium increased from 1980 to 1985. Especially after 1981, the differences between the water quality of the upstream water sources and those downstream (Iwakiri) increased markedly. So, too, did the levels of nitrate ion, calcium ion, chlorine ion, calcium, and magnesium, as did the pH of the water at the Iwakiri filtration plant. Thus, since 1981, when the Miyazaki-Oki plant was opened, pollution of the Iwakiri filtration plant has increased. This cannot be just a coincidence. In the winter of 1985, the citizens took a sample of drainage water at Kyushu-Komatsu and had it analyzed; 780 ppb trichloroethane was detected. This concentration was twice as much as the regulation level. We note that Miyazaki Oki (wafer process) and Kyushu-Komatsu (wafer manufacturing) are neighbors and that their drainage outlets flow together into the Kiyotake River. The Iwakiri filtration plant is located 1700 ft downstream (Fig. 4.6).

5.5.2 New Agreement for Environmental Protection

Neither the Agreement for Environmental Protection drawn up between Kiyotake City and Miyazaki Oki and Kyushu-Komatsu, nor the Agreement for Environmental Protection between Miyazaki City and Miyazaki-Oki has been made available to the public. In August 1984, the Environment Agency issued a notice about the regulation of trichloroethylene, and the agreements were subsequently revised, although not until 1986. The new agreements include items such as: (1) environmental protection plans; (2) chemicals; and (3) organic solvents. Nevertheless, these new agreements are insufficient to forestall high-tech pollution. This is because: (1) the agreements deal with only a number of the chemicals used by the semiconductor industry; (2) there is no limit set to the total amount of chemicals that may be used; and (3) the industries are not required to recover chemicals completely.

An IC lead frame manufacturer plans to locate its facility at Takanabe City, north of Miyazaki City, the site of many oyster farms. Because the facility uses plating, drainage treatment will be a big problem. Kiyotake City exempted Miyazaki-Oki from fixed property taxation for 3 years, and the city, therefore, lost an important tax revenue. This policy of attracting such famous companies to one's city raises many questions. In the future, it will be necessary for Miyazaki-Oki and Kyushu-Komatsu to disclose their environmental data, and for cities to set up monitoring systems. It is also necessary to reconsider the location of filtration plants beside rivers into which drainage from semiconductor plants flows.

5.6 Groundwater Pollution along the Tama River[41]

Because 60% of Tokyo's high-tech industries is concentrated along the Tama River, people call this locality "Tama High-Tech River". Residents near the Tama River use more groundwater than people in other areas in Tokyo. Here, too, the groundwater is polluted.

In October 1982, Fuchu City in Tokyo detected a high concentration of trichloroethylene (930 ppb) in the municipal wells. The Water Supply Division of Tokyo City had already ordered the closure of the wells. In December 1982, the Tokyo Environmental Protection Division disclosed its views about the nature and problems of the pollution as follows:[42]

1. Although we have investigated 18 plants which use trichloroethylene, we cannot locate the cause of the pollution.
2. We have also investigated T-electronics, Fuchu, which uses much trichloroethylene and where there is a contaminated well (774 ppb), but we cannot specify the direct cause of the pollution.
3. We have re-investigated four plants which dispose their drainage underground, As a result, we have finally found out that H-manufacturer used trichloroethylene before 1975. We also detected 1760 ppb of trichloroethane at the disposal box.
4. We cannot deny that H-manufacturer disposed trichloroethylene underground as a result of its working style. Further, the disposal of trichloroethane underground

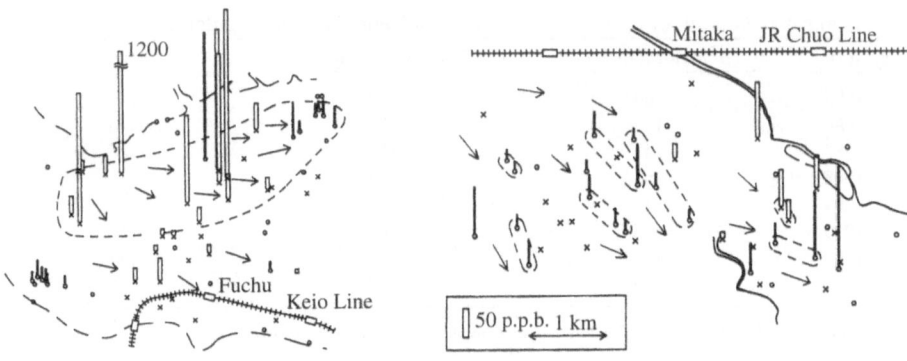

FIG. 4.7. Groundwater pollution in Tokyo (trichloroethylene)

is not advisable. We therefore have ordered H-manufacturer to stop disposing of trichloroethane and to improve its drainage facilities.

T-electronics, Fuchu, is Toshiba, Fuchu, which manufactures and distributes electronics parts, not semiconductors. According to the investigator, Toshiba had a special drainage system, but because its dry well had already been sealed off, the cause of the pollution was not established.

H-manufacturer is Hamai Manufacturer, sited next to the Toshiba plant, and it makes valves for LPG containers. We note that even at the time of the investigation (December 1982), Hamai was disposing trichloroethane underground, although the Fairchild case in Silicon Valley had already been reported and was known in Japan.

Later investigations have shown that the groundwater pollution at Fuchu is wider and deeper than at any other site in the Tokyo area. According to a simulation of groundwater pollution at Fuchu City by the National Environment Institute, ". . . the shallow well contamination has continuance", and ". . . it takes many years for the polluted groundwater to move even 0.6 mile".[43]

Another case of groundwater pollution, at Mitaka City, Tokyo, seems to have been caused ". . . by the direct disposal of solvent into old wells and dry wells", according to the investigator. Pollutants of organic solvent came from metal processing, printed circuit board manufacturing, automobile parts and dry cleaning (Fig. 4.7).

Groundwater pollution has also been caused by leakage from sewage plants, where electronics components factories have disposed solvents. For example, in the early 1970s, at Komae City, Tokyo, the disposal of organic solvent through the sewage by camera and electronics parts plants penetrated underground, and caused acute trichloroethylene poisoning of the workers at an underground sewage construction plant.

As early as 1974, the Tokyo Metropolitan Research Laboratory of Public Health had detected trichloroethylene and tetrachloroethylene in the groundwater. Groundwater pollution through sewage has also been recorded at a high-tech industry area near Boston, USA.

At Kawasaki City (Kanagawa Prefecture), on the opposite bank of the Tama River, the groundwater is also polluted. Kawasaki City houses many research and

TABLE 4.7. Amount of production of organic solvent (million pounds) in Japan

Year	Trichloroethylene	Tetrachloroethylene	Trichloroethane
1985	160	159	265
1986	157	154	281
1987	141	186	288
1988	154	213	305
1989	143	201	361
1990	125	184	408
1991	113	148	388
1992	135	139	369

Source: MITI (1985–1992) Statistical year book of chemical industry

development centers for the semiconductor industry. According to the Environmental Protection Division of Kawasaki City, "... the relationship between groundwater pollution and semiconductor related industries is not deniable, but it is difficult to specify the source of pollution" (*Asahi Shimbun*, Kanagawa Edition, October 27, 1988).

Many toxic chemicals have been detected in Kawasaki Bay. The Environment Agency has detected dioxin (2,3,7,8-TCDD) in the mud of Tokyo Bay (2 ppt). In 1986, Kawasaki City carried out an environmental survey of chemicals used at four semiconductor plants (NEC, 2 × Toshiba, and Fujitsu). The results, however, have not been made available to the public.

Since there are many possible sources of groundwater pollution along the Tama River, we can not necessarily call it "High-Tech Pollution". Nevertheless, the manufacture of electronics parts and printed circuit boards is related to the semiconductor industry, and Tokyo and Kanagawa Prefecture house many semiconductor research and development centers. In Kanagawa Prefecture, especially, where many high-tech industry are located, factories and private homes are found side-by-side. Recently, at Sagamihara City and Hadano City, where many high-tech factories are located, Kanagawa Prefecture has found high concentrations of contamination in the groundwater. According to a *Report of Specialty Gas Usage by Kanagawa Prefecture* (61 plants; 1985), the number of plants using toxic gases like silane and phosphine has increased threefold in 7 years. Many facilities have no special extinguishers or power sources for gas use. In 1984, there was an explosion of germane gas at Nihon Sanso, Kawasaki.

If we wish to fight high-tech pollution, we have to pay attention not only to groundwater contamination, but also to chemical pollution in general.

6. How to Protect Our Environment

6.1 Legal Regulation of Chemicals

6.1.1 Production of Chemicals

So serious and everyday is chemical pollution that PCBs have even been detected in a mother's milk. The present-day development and production of chemicals is so remarkable that the number of chemicals known to us has now reached eight million. Approximately 70 thousand chemicals are used as general commodities. Every year, approximately 1000 new chemical substances are generated.

One group of chemicals that has caused groundwater pollution is the organochlorines. Organochlorines have a relatively higher toxicity than other chemicals. The Japanese Chemical Substances Control Law designated nine substances as specified chemicals (the first category: PCB, PCN, HCB, DDT, aldrin, endrin, dieldrin, and chlordane). Recently, Japanese production of trichloroethane (a substitute for trichloroethylene) and tetrachloroethylene (a substance in CFCs) has increased; so, too, has dichlorobenzene (used as a pesticide). Organochlorine was originally produced to assist the disposal of chlorine as a by-product of the electrolysis of sodium hydroxide.

These chemicals penetrate the environment in many ways. *The Report of the Workshop on Practical Approaches for the Assessment of Environmental Exposure* published by the OECD, Environment Directorate, has listed the sources of exposure as follows:[44] (1) point discharge to water from industrial sources; (2) release to water from municipal sewage treatment plants; (3) point source release to air; (4) release to air from disperse source; (5) direct deliberate application to soil; (6) landfill disposal of wastes; and (7) incineration of wastes.

6.1.2 Revised Japanese Chemical Substances Control Law[45]

Although the Japanese government has now passed many laws to regulate pollution caused by the high-tech industry, these laws lack uniformity. In April 1987, the Chemical Substances Control Law (The Law Concerning the Examination and Regulation of Manufacture, etc., of Chemical Substances) was amended to cope with chemicals involved in high-tech pollution. This amended law requires the classification of a substance that is suspected of biological degradability and toxicity as a "specified chemical substance". If the harmfulness of this substance is confirmed, it is classified as "a second category of specified chemical substances", the production and import of which must then be regulated. The Ministry of International Trade and Industry is mainly in charge of testing substances for biologic degradability and the accumulation of chemicals in fish, while the Ministry of Health and Welfare tests their toxicity through the examination of animals. At the same time, the Environment Agency carries out studies and surveys of the state of chemical substances in the environment. The most important problem raised by the amended chemical substances control law is that even if a "specified chemical substance" is suspected of toxicity, its discharge into the environment is allowed if the chemical is not dispersed over a "very wide area". Yet, groundwater pollution by organic solvent is dangerous to human health, irrespective of its accumulation or dispersal. The amended law is not sufficiently stringent to cope with groundwater pollution, which was what triggered the revision in the first place. The environmental investigation of chemical substances by the Environment Agency detected 28 of 82 chemicals that the Ministry of International Trade and Industry has checked for "degradability" in their effect on the environment (Fig. 4.8).

Because applications to use chemicals are made to both the Ministry of International Trade and Industry and the Ministry of Health and Welfare, chemical data are

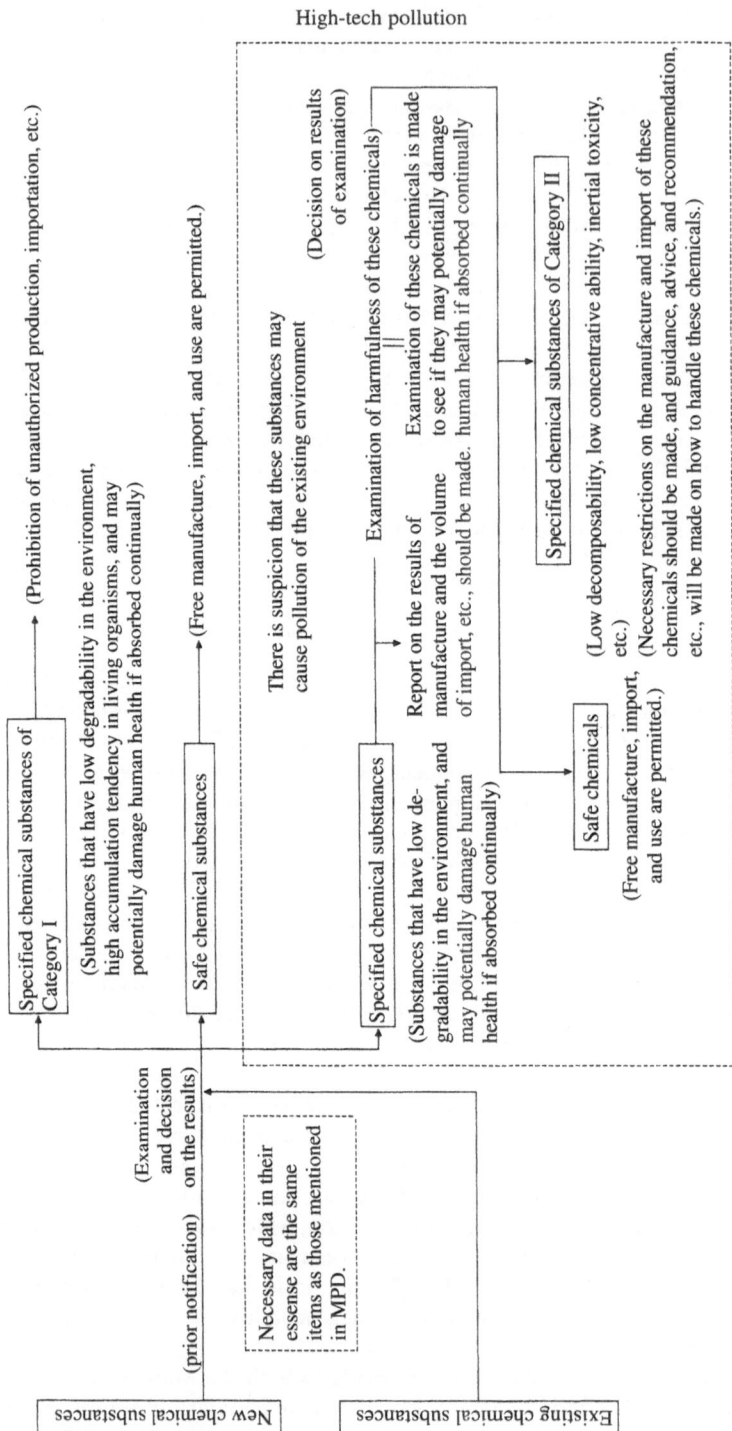

FIG. 4.8. Regulatory system under the Chemical Substances Control Law. [ ]; parts of major amendment made in the regulatory system under the Chemical Substances Control Law. *MPD* (Minimum Premarketing Set of Data) means those items for assessment of the minimum safety of chemicals before they are put on the market, which were recommended by the Council of the OECD in 1982.
Source: Environment Agency, Government of Japan (1990) Quality of the environment in Japan, 1989. Printing Bureau, Ministry of Finance, Tokyo, p 112

gathered by both ministries. Such information should be made open to public inspection. In Japan, however, the results of the toxicity testing of pesticides registered at the Ministry of Agriculture and Fishery is not disclosed, because the data are regarded as the "property of the company".

In the USA, however, the Emergency Planning and Community Right-to-Know Act of 1986 (SARA Title III) is a very good guide, because it requires the owner of chemicals to open to the community knowledge of: (1) emergency planning; (2) a material safety data sheet; (3) a list of extremely hazardous substances; and (4) an emergency and hazardous chemical inventory.

Groundwater pollution at Kimitsu City triggered the revision of the water pollution control law in 1989: (1) organic solvents like trichloroethylene are included among substances subject to the control of law; and (2) penal regulations are added to clauses relating to the protection of groundwater.

Although the Ministry of International Trade and Industry designated trichloroethylene, tetrachloroethylene, and carbon tetrachloride as falling within "the second category of a specified chemical substance", trichloroethane is not included in the list (Table 4.7). In 1988, the Ministry of Labor issued a guideline for toxic gases used at semiconductor plants: (1) facilities for gas supply should be made of suitable material and be constructed to prevent gas leakage; (2) clean rooms should have adequate ventilation systems, ventilation times and emergency exits; (3) every piece of apparatus should have equipment designed to prevent gas leakage and electric shock; and (4) during maintenance work, special attention should be paid to the control of toxic gas. This guideline was based on a survey of 76 plants and the issue of the guideline itself is appreciated. Nevertheless, the results of the survey are not open to the public and the revised safety standards for the use of toxic gas are not sufficiently clear.

As to organic solvent, the Ministry of Labor has revised the Occupational Safety and Health Law and designated a total of 47 organic solvents, 35 of which need to be labeled as toxic. The law also requires the examination of the workers' environment.

6.2 An Agreement for Environmental Protection That Requires Manufacturers to Submit Names of All Chemicals Used—Kitakami City and Iwate-Toshiba-Electronics

6.2.1 Iwate Toshiba-Electronics

Kitakami City is located along the Kitakami River in Iwate Prefecture in the northeastern part of Japan. Iwate-Toshiba-Electronics in Kitakami City has attracted the attention of other local governments because Toshiba has signed an agreement with Kitakami City for environmental protection; this includes the naming of chemical items used.

Iwate-Toshiba-Electronics (2500 employees), which opened in 1973, began making semiconductors in 1984. Because Toshiba is within easy access of the Tohoku Highway and is able to tap abundant water and recruit a workforce, Iwate-Toshiba-Electronics has become a center of ASIC production in eastern Japan. Because of the poor quality of the groundwater, Toshiba uses industrial water from the Kitakami River (2.6 million gallons per day).

6.2.2 The Agreement for Environmental Protection

The most significant aspect of the agreement for environmental protection between Kitakami City and Toshiba is that the second item orders Toshiba to present data on the chemicals used at the plant at the request of the Kitakami City Authorities. The fifth item stipulates the prevention of groundwater pollution, while the 10th item authorizes the City's inspection of the plant. These items are appropriate for application to all semiconductor plants. Once Kitakami City had proposed to draw up agreements for environmental protection with 18 companies at the Industrial Park, the City concentrated on Toshiba's semiconductor plant, and a task force from the environmental protection committee of Kitakami City subsequently carried out research and a survey of Toshiba, Taishi, and presented a draft of an agreement to the Mayor of Kitakami.

6.2.3 Subject for Future Study

The actual agreement differs in significant ways from the draft: (1) what the draft defined as "measure and record" the emission of substances that are not under the control of the law becomes "separately consult"; (2) the draft proposed an "entrusted specialist" as a "member of the inspection committee", but the agreement deletes this item; and (3) the agreement introduces a "trade secret" item (item 13). In fact, Kitakami City did not disclose the draft of the agreement, but kept it secret too.

The chairman of the task force, Professor Tatsuo Goto (Iwate University), says that because the item "separately consult" has the character of a "watching" item, the contents of the agreement itself will be decided by the power relationship between Kitakami City and Toshiba. Later on, in September 1987, the parties agreed on a memorandum that Toshiba will independently report the results of measurements of 22 toxic chemicals (drainage once per month or per 4 months; air twice per year). Among 22 toxic chemicals that are unregulated by laws, six substances have to be drained (phosphorus, nitrogen, boron, silicon, carbon tetrachloride, and xylene) and seven substances are emitted as exhaust (fluorine, hydrogen chloride, phosphorus, boron, carbon tetrachloride, xylene, and arsenic). The city itself can inspect organic solvent. Toshiba has already changed over to CFCs. Although the city had carried out three inspections by September 1987, the standards of regulation and conditions for waste disposal trading were still not clear.

Iwate-Toshiba-Electronics started the production of charge-coupled devices (CCD) in 1987 and ASIC in 1988. In 1990, the production level doubled, while the numbers of part time jobs had been increasing. Therefore, the building up of a "watching system" for the environment and strict safety training systems are required now more than ever.

Local governments should employ special staff responsible for watching high-tech "secrets", and the national government should strengthen its regulations on toxic chemicals.

6.3 An Agreement for Environmental Protection Including a "Closed Drainage System" and "Toxic Gas Regulation"— Tateyama City and NMB Semiconductor (Japan Foundry)

6.3.1 Location of the Semiconductor Plant

Tateyama, located at the edge of the Bohsoh Peninsula, is an agricultural and fishery area, with few factories. However, in April 1984, a plan for a semiconductor plant was proposed. At that time, because of the boom in the semiconductor business, NMB, a manufacturer of ball bearings, was searching for a site for a semiconductor plant. NMB planned to make 256 k DRAM under license from Inmos (UK).

Because Tateyama is near Tokyo and has a good environment, NMB chose Tateyama as its candidate location. In addition, because Chiba Prefecture was particularly eager to attract high-tech industries, they confirmed approval of the plan within only 3 months (August 1984).

The most difficult issue, however, has turned out to be the pumping up of groundwater. Since the groundwater dried up as a result of military use at the time of World War II, this area has no large groundwater basin. In addition, the municipal well and an agricultural pond are located near the site of the plant. Citizens living near the site therefore opposed the plan, and NMB nearly gave up the idea.

Because, at the same time, "High-tech pollution in Silicon Valley" was also being reported in Japan, the City Council discussed this problem and strict regulations were required.

6.3.2 Contents of Agreement for Environmental Protection

In Japan, today, approximately 220 agreements for environmental protection between local governments and semiconductor plants, and research and development centers have been drawn up. However, only 10% of these agreements deal with chemicals that are not already regulated by law. Among them, the agreement for environmental protection between Tateyama City and NMB semiconductor (January 1985) may act as a model, because it includes a "closed drainage system" and a "toxic gas regulation".

This agreement was based on an agreement drawn up between Miho village (Ibaragi Prefecture) and Texas Instruments, Japan. The amount of groundwater pumped up was reduced from 800 000 gallons per day to 160 000 gallons. A "closed drainage system", which uses water only for miscellaneous purposes and evaporation, was adopted. Waste chemicals have to be extracted and disposed of as industrial waste.

Although a "closed system" (Organo Corporation) costs ¥4500 million (approximately US$30 million; total investment ¥30 000 million, approximately US$200 million), it is necessary to take special measures to protect the environment.

Twenty-two toxic chemicals have emission standards based on those set by the American Conference of Governmental Industrial Hygienists (ACGIH). A "watching system", however, is only self-monitoring of the plant by the plant (Table 4.8).

Recently, the NMB semiconductor (now Japan Foundry) company started the production of 4 Mega DRAM and pumped up the maximum limit of groundwater that had been modified to 320 000 gallons per day. It thus becomes necessary for the citizens and city to set up a regular and continuous "watching system" (Fig. 4.9, Fig. 4.10).

TABLE 4.8. Environmental protection plan of NMB Semiconductor (1985)

Industrial waste

	Source of generation	Form	Amount/year
Sludge	Water treatment	Sludge	2.6 million pounds
Waste plastics	Production process	Solid	53 000 pounds
Sulfuric acid	Resist	Dense liquid	904 000 pounds
Nitric acid	Diffusion	Dense liquid	102 000 pounds
Ammonia	Diffusion	Dense liquid	33 000 pounds
Fluohydric acid	Etching	Dense liquid	550 000 pounds
Organic solvent	Resist	Liquid	64 000 pounds
Developer	Resist	Liquid	260 000 pounds

Amount of use

	Process	Amount/month
Sulfuric acid	Etching	5 000 gallons
White phosphorus	Ion implantation	1 100 pounds
Ammonia	CVD	13 359 gallons
Hydrogen fluoride	Etching	1 900 gallons
Acetone		666 pounds
Isopropyl alcohol		444 pounds
Metallic arsenic	Ion implantation	8.8 ounces

CVD, chemical vapor deposition

The example of Tateyama City shows that because the city and citizens did not originally want to attract a semiconductor plant, they have been able to take a strong position and insist on strict regulations. We must, therefore, once again question the regional development policy of attracting plants to rural or apparently suitable sites.

6.4 To Forestall High-Tech Pollution[46,47]

When I look at a high-tech plant operation from the backyard, I see that high-tech facilities use and store many sorts of toxic chemicals and gases in on-site storage tanks and cylinders. This is my impression during field research of high-tech pollution in Japan and the USA.

When I visit such places, I notice that many sorts of high-tech pollution have occurred and agreements for environmental protection between cities and semiconductor plants have been drawn up. We have to pay attention to what they express in common, rather than their variety.

First, an uncontrolled high-tech industry is an essentially dangerous process. We should note, too, that factories that manufacture watches, commutator, and electric automobile parts, which, at first sight, are not related to high-technology, sometimes fabricate semiconductors and, therefore, use organic solvents and toxic chemicals. If local governments and citizens plan to attract factories, it is indispensable to know what these factories manufacture and how the plant will be used. Needless to say, it is not only necessary for the local government to draw up an agreement for environmental protection, but also for the agreement to include specific regulations regarding toxic chemicals. In cases where regional development policy aims to attract

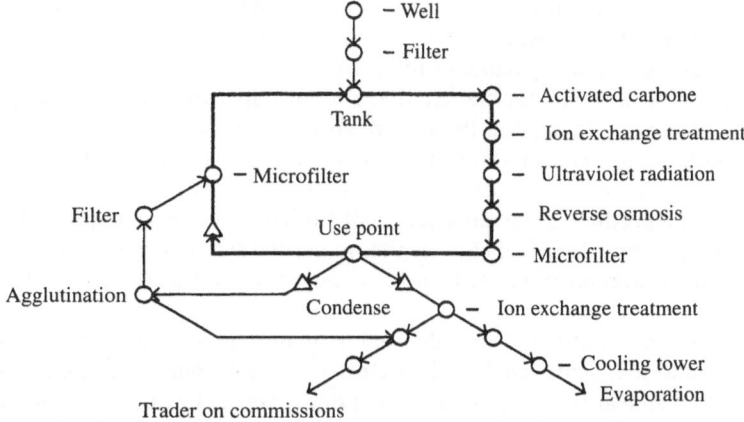

FIG. 4.9. Closed drainage system at NMB Semiconductor

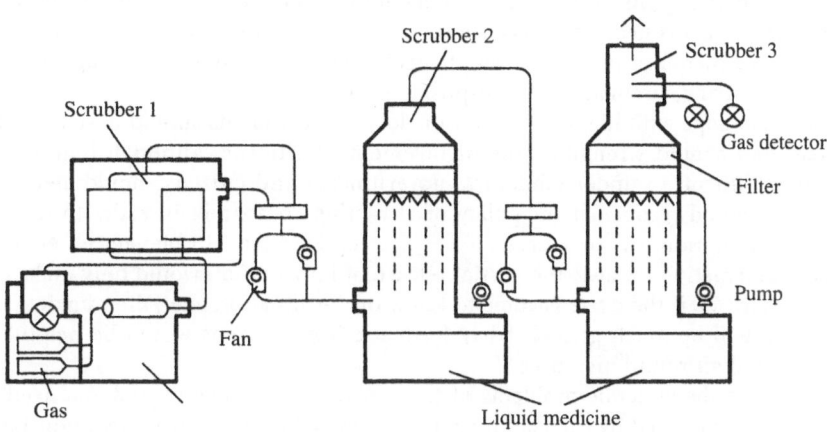

FIG. 4.10. Process of Gas treatment

factories, local governments and the citizens tend to be submissive. Now that, as a last resort to vitalize local economies, high-tech industry is welcomed from all quarters, this need becomes ever more urgent.

Second, groundwater pollution and the disposal of toxic chemicals have reached an emergency level. In Japan, despite the continuation of groundwater contamination, the analysis of causes and routes is too protracted. In the USA, before the cleanup begins, much time and money is spent on research into the causes and routes of pollution. If the investigation is inadequate, pollution will only grow worse, and the delay in taking action will cause even greater damage and even higher costs.

Japan has many legal problems to sort out. In Japan there are only 11 substances that are regarded as harmful to human health (CN, R-Hg, Or-P, Cd, Pb, Cr, As, Hg, PCB, trichloroethylene, and tetrachloroethylene). Even the Basel Convention on the control of transboundary movement of hazardous waste and their disposal designated only 47 substances. On the other hand, in the USA, SARA has listed 406 hazardous substances (1988), while CERCLA has listed 721 hazardous substances (1988). Japanese regulations on waste disposal in the sea are also inadequate. Nor are regulations on organic solvents and CFCs anything like strict enough. If we fail to control each chemical until "accumulation" in the environment has been confirmed, or if we say that the suspicion that a product is carcinogenic is not sufficient reason to regulate its use, we are, in fact, approving "experiments on the human body" on a very large scale. It is high time for us to remember the lessons we ought to have learned from Minamata disease.

Third, information about chemicals used as materials or industrial commodities should be collected, centralized, and disclosed to the public. Citizens are anxious about what is done at factories and how it is done. Yet, as businesses themselves are so nervous about "high-tech", they try to keep the details secret, which is common sense in a ruthless business world. In this respect, the USA legal system and the citizens movement of the "Right-to-know" offer a model. At the same time, the government should change its policy that ". . . data about chemicals are the property of the companies". This is necessary because the use of these chemicals mostly affects the public. It is a matter of urgency to set up legal systems that control the whole process of manufacturing, circulation, and disposal of industrial chemicals.

Fourth, to cope with high-tech pollution, local governments should have a right to "watch" and should strengthen the involvement of citizens. Although high-tech is somehow difficult to understand, local governments and citizens should, nevertheless, analyze and reconsider the policy of attracting companies in order to forestall high-tech pollution. In some cases, local governments tend to blanket all environmental information because the partial release of information would only make citizens uneasy. Yet, if the citizens come to know of the situation later, their anxiety and confusion will be much greater. Early feedback from citizens would be helpful for taking initial all-round measures.

Because of the endemic problems of trade friction, the safety standards fixed for commodities imposed by each country have been criticized as forming a non-tariff-barrier. Because of the competition between countries, information about chemicals tends to be kept secret and investment on safety is thus saved. Confrontation between nations then tends to be stirred up. It is high time that we move beyond national borders and the restrictions imposed by individual companies, and all citizens of the world should be able to exchange information, especially where it concerns safety and health.

References

1. Epidemiological Studies Section, California Department of Health Services (1985) Pregnancy outcomes in Santa Clara County 1980–1982. Berkeley
2. Dapson SC et al. (1984) Effect of methyl chloroform on cardiovascular development in rats. Teratology 29:25A
3. Epidemiological Studies and Surveillance Section, California Department of Health Services (1988) Pregnancy outcomes in Santa Clara County 1980–1985. Berkeley

4. Lagakos SW et al. (1986) An analysis of contaminated well water and health effects in Woburn, Massachusetts. J Am Stat Assoc 81:583–596
5. Byers VS et al. (1988) Association between clinical symptoms and lymphocyte abnormalities in a population with chronic domestic exposure to industrial solvent contaminated domestic water supply and a high incidence of leukaemia. Cancer Immunol Immunother 27:77–81
6. National Institute of Environmental Health Sciences (1987) National toxicology program, developmental toxicity evaluation of 1,1,1-trichloroethane administered to CD rats. Final report. NTP-87-220. Research Triangle Park
7. Wrensch M et al. (1990) Hydrogeologic assessment of exposure to solvent-contaminated drinking water: pregnancy outcomes in relation to exposure. Arch Environ Health 45:210–216
8. Swan SH et al. (1992) Is drinking water related to spontaneous abortion? re-viewing the evidence from the California Department of Health Services studies. Epidemiology 3:83–93
9. State of California, Department of Industrial Relations, Division of Occupational Safety and Health, Task Force on the Electronics Industry (1981) Semiconductor industry study. Sacramento
10. US Environmental Protection Agency (1987) Santa Clara Valley integrated environmental management project, stage two report. San Francisco
11. US 99th Congress, First Session (1985) Hearing before the Subcommittee on Investigations and Oversight of the Committee on Public Works and Transportation, House of Representatives. San Jose. GPO, Washington DC
12. Patrick R et al. (1987) Groundwater contamination in the United States, 2nd edn. University of Pennsylvania Press, Philadelphia
13. State of California, op cit.
14. Environment Agency et al. (1987) A survey of the IC industry and environmental protection. Tokyo, p 160
15. Gassert T (1985) Health hazards in electronics. A handbook. Asia Monitor Resource Center, Hong Kong
16. LaDou J (ed.) (1986) Occupational medicine, the microelectronics industry, state of the art review, Vol 1, No 1. Hanley and Belfus, Philadelphia
17. Robbins P et al. (1989) Summary of occupational injuries and illnesses in the semiconductor industry for 1980–1985. In: American Conference of Government Industrial Hygienist (ed.) Hazard assessment and control technology in semiconductor manufacturing. Lewis, Chelsea, pp 3–16
18. Pastides H et al. (1988) Spontaneous abortion and general illness symptoms among semiconductor manufacturers. J Occup Med 30:543–551
19. State of California, op cit.
20. Environment Agency et al., op cit., p 103
21. Lin VK (1986) Health, women's work and industrialization: semiconductor workers in Singapore and Malaysia. PhD thesis, University of California, Berkeley
22. Division of Occupational/Environmental Medicine and Epidemiology, University of California at Davis (1992) Epidemiologic study of reproductive and other health effects among workers employed in the manufacture of semiconductors, final report to the Semiconductor Industry Association. Davis
23. Correa A (1996) Ethylene glycol ethers and risks of spontaneous abortion and subfertility. Am J Epidemiol 143:707–717
24. Schenker M et al. (1995) Association of spontaneous abortion and other reproductive effects with work in the semiconductor industry. Am J Ind Med 28:639–659
25. Kizer KW et al. (1988) Sound science in the implementation of public policy. JAMA 260:951–955

26. Pease WS (1988) Environmental pollution and cancer in California: evaluating the significance of risks under proposition 65. Master's thesis, University of California, Berkeley
27. Siegel L et al. (1985) The high cost of high tech. Harper and Row, New York
28. US Environmental Protection Agency, op cit., pp 3-69-3-73
29. US Environmental Protection Agency (1987). Waste minimization, environmental quality with economic benefits. Washington DC
30. US Environmental Protection Agency (1988) The citizens' guidance manual for the technical assistance grant program. Washington DC
31. Association of Industry and Environment (1986) Report of high-tech industry and environment [in Japanese]. Tokyo
32. KHN (Society for Safety of Condensed Gas) (1987) Reference book for safety of specialty gas [in Japanese]. Tokyo
33. State of California, op cit., p 13
34. Science Forum (1987) Manual book for safety of high-technology [in Japanese]. Tokyo
35. Webb D (1984) In vitro solubility and in vivo toxicity of gallium arsenide. Toxicology and Applied Pharmacology 76:96–104
36. Yoshida F (1980) The economics of environment and technology. Aoki Shoten, Tokyo
37. Nakasugi O (1984) Groundwater pollution by organochloride [in Japanese]. *Kikan Kankyo Kenkyu* [Q J Environ Study] 52:125–135
38. Yamada K (1985) Warning from groundwater (2) [in Japanese]. *Gijyutsu to Ningen* [Technol Man] April: 10–26
39. Kobayashi T (1986) Example of Hyogo Prefecture. In: Japan Water Pollution Research Association groundwater safety research, case study. pp 149–183
40. Suzuki S (1988) Groundwater pollution in Kumamoto City [in Japanese]. *Sangyo Keiei Kenkyu* [Study Industry Economy] 6:37–60
41. Morita M et al. (1974) An investigation analytical method and evaluation of organic matters. In: Annual report of Tokyo Metropolitan Research Laboratory of Public Health [in Japanese], Vol 25. Tokyo, pp 399–403
42. The Tokyo Watch Committee of Pollution (1983) Bulletin No 73. Tokyo
43. Muraoka H (1986) A simulation of groundwater pollution by organic solvent. In: Basic research on mechanism of groundwater pollution by organic solvent. National Institute, Tsukuba, p 77
44. OECD Environment Directorate (1986) The report of the workshop on practical approaches for the assessment of environmental exposure. Paris
45. Environment Agency (1987) Study of basic information about toxic chemicals [in Japanese]. Tokyo
46. Oikawa K (1987) Profile 100 of hazardous and toxic chemicals in high tech industry [in Japanese]. Maruzen, Tokyo
47. Kinki Legal Profession (1988) Dangerous groundwater pollution [in Japanese]. Osaka

Chapter 5
The Current State of High-Tech
Pollution*

1. Introduction

Japan has waited a long time for the implementation of a cleanup system that will deal with "groundwater pollution".

The Water Pollution Control Law (1989) was partially revised at the end of May 1996 and, from April 1997, in cases where there is danger that human health will be affected by hazardous material polluting the groundwater, the governor of the prefecture or the ordinance-designated mayor is empowered to order the polluter to take retroactive water quality purification measures. Although revision of the law should have led to progress in water purification, many problems remain unsolved.

The first thing we have to do is to investigate the mechanisms that cause groundwater pollution and then clarify who is responsible for cleanup. Social needs for groundwater, which vary from prefecture to prefecture, will determine what has to be done to prevent the pollution as well as the eagerness with which the interested parties deal with the cause of the pollution.

The second issue is one of solvency. This has to do with whether or not the polluter can pay for the cleanup expenses, including the cost of the initial investigation. One question is whether the local government body should have to bear the cost in cases where the source of the pollution is unspecified or when the specified source is an insolvent small enterprise or plant to which the designated facility "cannot be applied".

Consequently, I should like to examine, in some detail, several characteristic cases of high-tech pollution currently under investigation as examples of the Japanese contribution to ways of dealing with geo-pollution.

2. A Model of "High-Tech Pollution" Cleanup: Kimitsu City

The Toshiba Component Kimitsu Plant, located in the Uchi-minowa area of Kimitsu City, Chiba Prefecture, employs 500 workers and produces commutator semiconduc-

---

* Originally published in *Economic Journal of Hokkaido University* (Sapporo, 1997) 26:21–44. With the permission of the University.

tor silicon chips. Trichloroethylene is used in the manufacturing process and, in 1987, it was found that trichloroethylene had seeped into and polluted the groundwater. Yet, it was not until September 1988, a year and a half after groundwater pollution had been discovered at the plant, that the public was finally informed of the discovery.

A formal investigation was conducted by Kimitsu City and Chiba Geo-Environment Research Laboratory, the first detailed investigation of pollution of this kind to be carried out in Japan. On the basis of data produced by the investigation, the cleanup of the polluted groundwater was dealt with.

The Kimitsu investigation and cleanup was characteristic of what happens when appropriate measures of selection, development, and enforcement are taken by geological specialists in the project team to tackle the mechanism of pollution. The measures taken were based on the concept of geo-pollution, which covers issues such as strata, groundwater, and ground air pollution.[1] The pollution source was determined after investigation of the surface layer by methods the investigators developed on the spot.

The investigators found seven highly polluted "hot-spots", the causal features of which were revealed by a time series of aerial photographs and the employee's report: (1) dumping into an underground waste dumping site near the fifth well; (2) leakage during the process of refueling the trichloroethylene tank (First Production Section); (3) leakage during the process of refueling the trichloroethylene tank (Second Production Section); (4) leakage during the process of refueling the trichloroethylene tank (Thirty-sixth Building); (5) leakages when replacing waste liquid in front of the cafeteria; (6) leakages while carrying waste liquid along the passage from the Second Production Section to the cafeteria; and (7) dumping of trichloroethylene used for cleaning clothes in the passage beside the warehouse.

Investigation of 17 borings at nine places within the plant site and of 17 borings at 17 places within the city area revealed the distribution of the aquifer and the actual extent of the geo-pollution. At the same time, a corresponding cleanup technology was developed.

Subsequently, investigators found a dual mechanism contributing to the contamination: leakage from surface facilities at the building site, and contamination of the underground strata (Fig. 5.1).

First, waste substances and contaminated strata were removed, after which heating and air-drying treatments were undertaken on contaminated strata. To deal with groundwater pollution, pumping and aeration treatments were carried out in each catchment basin well, whereas the sucking method of vaporization and capture of trichloroethylene was adopted for dealing with ground air pollution.

Second, a shield of steel tubing was put in place at the factory site and the barrier well system was adopted lest contaminated substances from the plant be diffused within the city area. The contaminated substances and groundwater were effectively discharged from the aquifer.

Third, the cleanup operators were encouraged to pump, aerate, and use water from public wells in order to remove contaminated substances diffused within the aquifer of the city area. The facilities of Uchi-minowa Sports Park are representative of the usefulness of public wells. These cleanup measures helped considerably to reduce the concentration of trichloroethylene (in ppb) in the groundwater to a three-digit number within the plant premises, to a two-digit number near the barrier well, and to approximately 10-fold the water quality criteria in the city area compared with the situation before the cleanup.

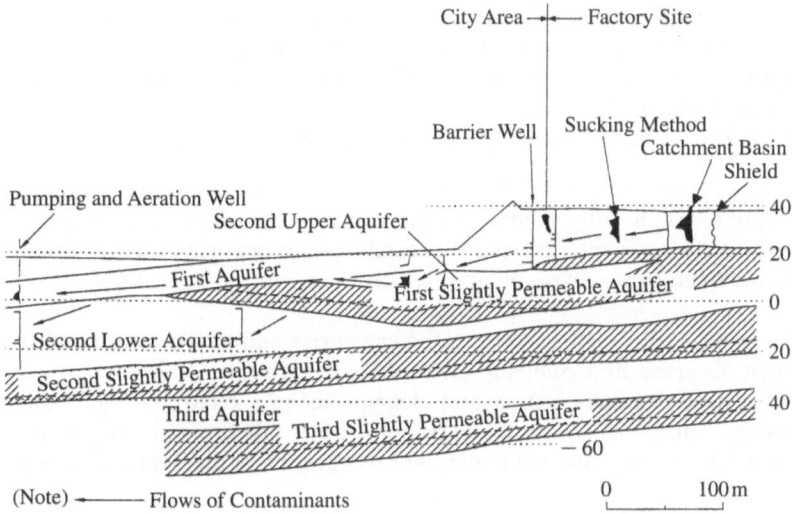

FIG. 5.1. The geo-pollution and cleanup technology at Kimitsu
Source: Department of Environment, Kimitsu City (1993) The first report of cleanup measures
of geo-pollution

The Kimitsu experience is extremely important because, for the first time, information about groundwater pollution was made public in response to the citizens' uneasiness and their criticism that information had been kept secret for approximately a year and a half after the initial discovery of groundwater pollution. Although the total cost of the investigation and cleanup has so far amounted to less than ¥1.2 billion, including approximately ¥50 million borne by the city, the PPP has been fundamentally adhered to.

The estimated costs for removing the source of the pollution are approximately ¥0.2 billion for installing the steel tubing, digging, transportation, back filling, and installation of a catchment basin and triple aerator, and an additional cost of approximately ¥60 million for air-drying equipment and so on, all of which are to be paid for by the polluter. As to provision of the barrier well system, the polluter bore costs of approximately ¥0.1 billion for the installation of the system, and approximately ¥3 million for annual operating costs. In addition, the cost of the equipment for pumping, discharging, and aeration, approximately ¥7.8 million, was to be paid by the polluter, whereas the annual operating cost, approximately ¥8 million, was to be partially paid by the polluter.

Medical examinations of the residents have continued but, so far, no abnormalities caused by groundwater pollution have been reported. Toshiba Component Company, negotiating with the Uchi-minowa Groundwater Pollution Countermeasure Committee representing the residents of approximately 300 households (i.e., approximately 1000 people), agreed to pay compensation of ¥37 million in April 1992. This figure does not, however, include compensation for any future damage to health.

The Kimitsu groundwater pollution case prompted amendment of the Water Pollution Control Law enacted in 1989, and the resultant regulation of groundwater has had a great effect throughout Japan. The discovery of the mechanism by which this pollution occurs and the resultant cleanup technology, which was developed by Kimitsu, have subsequently been made public, so that people who want to learn about the issue may do so.

Kimitsu was able to make such a thorough investigation because:

1. Groundwater, including the source of city tap water, was widely polluted;

2. Nippon Steel Kimitsu Plant is located nearby, where the administration had an insight into environmental issues, and excellent technical support was provided by staff of the Geological Environment Research Center of the Prefectural Government; and

3. In terms of the relation between the enterprise and the residents, which differs from that observed in a company town, the enterprise was able to take a positive attitude toward the investigation and cleanup, using a relatively high standard of technology. The prefectural administration, the residents, and the enterprise all cooperated in finding a solution under joint recognition of the importance of groundwater resources.

3. What Has Happened at the Toshiba Taishi Plant Since High-Tech Pollution Was Discovered in Japan for the First Time?

Taishi City, Hyogo Prefecture, is recognized as the first known site in Japan of groundwater pollution caused by a semiconductor plant. Taishi City is located west of Himeji, is convenient for transport, and has a growing population. Taishi City is also the home of Toshiba's Manufacturing Plant. Groundwater, including the tap water supply source, was contaminated over a large area by trichloroethylene. The pollution source was the Toshiba Taishi Plant, formally the Himeji Semiconductor Plant, Ibo District, where trichloroethylene was used for the cleaning of semiconductors and cathode ray tubes (CRT).

Upon investigation, soil contamination was found to be caused by problems arising from the "... storage and the method of use" of trichloroethylene, stored near the south side of the 407th building of the plant. The digging out of contaminated soil was interrupted by the gushing of groundwater at a depth of 7 m. The concentration of trichloroethylene in the plant's shallow well rapidly reduced the quality of tap water, whereas the water in the deep wells retained a high concentration of trichloroethylene of approximately 8000 ppb. Similarly, cis-1,2-dichloroethylene, more poisonous than trichloroethylene, was also detected in nearby wells. In 1990, the Japan Water Pollution Research Association analyzed the site and found that contamination of the deep wells of the plant with high concentrations of trichloroethylene, caused by elution of trichloroethylene as a result of boring, was present in the soil 35–55 m under ground, and that this contamination was eluting further into deeper groundwater.[2] Due to its low solubility in water, trichloroethylene continued to elute, hardly lowering the degree of contamination.

Although the use of trichloroethylene at the plant was halted, trichloroethylene remaining in the soil continued to pollute the shallow well. It was thus necessary to

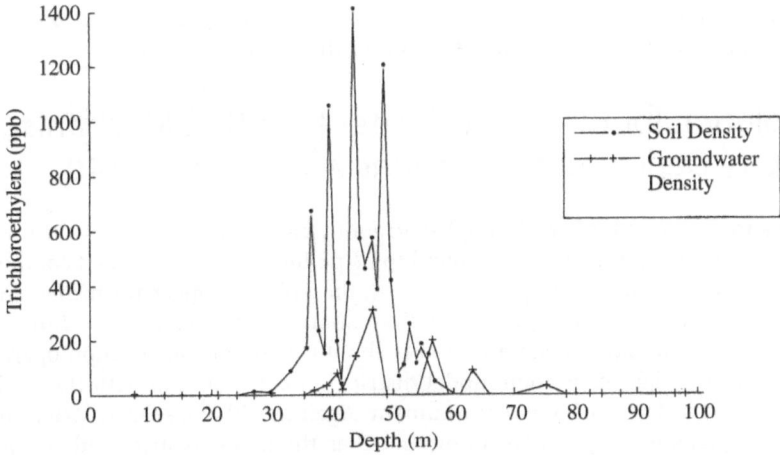

FIG. 5.2. The density of trichloroethylene at Toshiba Taishi Plant
Source: Japan Water Pollution Research Association (1990) Groundwater safety research, case study. A report to Environment Agency, Tokyo, p 102

determine exactly the place where the trichloroethylene was retained. In some nearby wells, cis-1,2,-dichloroethylene was detected at a higher concentration than trichloroethylene. This is because cis-1,2,-dichloroethylene originates from the microbial degradation of trichloroethylene. For a thorough solution to the problem of "contaminated soil", the contaminated soil should, in my view, be removed during the early stages of the investigation and we must recognize the problems caused when trichloroethylene has been stored underground for a long time. Inadequate solutions merely lengthen the duration of the pollution. In the case of the Fairchild Semiconductor Company in Silicon Valley in the US, the soil was dug out to a depth of 40 m at the source of the pollution. The first thing necessary to clarify the pollution mechanism is to investigate soil gas in order to trace the pollution source in shallow wells, such as those at Kimitsu. Hence, we have to return to basic principles. cis-1,2,-Dichloroethylene is not formed until groundwater is thoroughly polluted, which emphasizes the danger of neglecting the first signs of groundwater pollution (Fig. 5.2).

After the discovery of groundwater pollution in wells used as sources for tap water in Taishi, aeration and activated carbon measures were taken, and the water supply works were equipped to enable a change from private wells to tap water, but some wells were still used as sources of water for showers and baths.

Toshiba Taishi Plant refused to accept official responsibility for the pollution, and persisted in calling their contribution to the cost of converting water sources from wells to tap water supply a "donation" rather than "compensation". The plant substituted 1,1,1-trichloroethane for trichloroethylene. Even the response of Toshiba's subsidiary plants is quite different from that of Toshiba at Kimitsu; this is mainly a reflection of whether the local government takes strong action or not. Near Hitachi's Sakura Plant in Sakura City, Chiba Prefecture, which produces the same CRT as Toshiba's Taishi Plant, trichloroethylene (170 ppb) was detected in the spring water on the town's outskirts in 1990, after which pollution of the first to the third aquifers was discovered by geo-investigation of the downstream groundwater flow.[3]

4. Solution for the Groundwater Pollution Mechanism of a City Dependent on Groundwater: Kumamoto

Kumamoto City, Kyushu Island, supplies groundwater to all households in the city. Ten years ago, the groundwater was found to be polluted by a chlorinated compound. Although the city water was widely polluted, the pollution mechanism has not yet been fully clarified. In order to analyze the pollution mechanism, both Kumamoto Local Government and Kumamoto City set about a full-scale purification operation, establishing a "pollution cleanup model district" in collaboration with the National Environment Institute of the Environment Agency. Takahiradai District in the north of Kumamoto City has been nominated as the model district and consists of a terrace 80 m above sea level, with alluvial low land 20 m above sea level; it is a semi-industrial area, equipped with tap water from a well supply source on the city's outskirts.

In 1987, trichloroethylene was first detected in Takahiradai District, with concentrations exceeding the standard value, but the cause of the contamination was then unknown. From 1990 to 1991, Kumamoto City Government undertook an investigation of surface gas and found highly concentrated trichloroethylene soil gas in the building and parking lots of Kyushu Electric Industry, which had previously used trichloroethylene, as well as in the neighboring company's parking lot. As a result of executing 14 borings, the investigators found more than 10 mg/kg trichloroethylene on the premises, at depths of between 20 and 60 m and over an area 30 m in diameter. Pollution was found to be distributed underground in an onion-shaped pattern around the site of highest concentration, and this was regarded as the groundwater pollution source.[4]

As in the case of Kimitsu Toshiba Component, a dual mechanism responsible for the pollution was discovered in Kumamoto City too. As much as approximately 27 tons trichloroethylene had been used by Kyushu Electric Industry for cleaning the points of telephone dials over 10 years from 1967.[5] The trichloroethylene contaminating the soil was collected by pumping and by the gas-absorption cleanup method; however, since 1994, only the pumping method has been used.

Because this was a model case, the total cost of groundwater pollution abatement, approximately ¥450 million, was shared by the Kumamoto local government, Kumamoto City, and the central government. At first, Kumamoto City considered ordering Kyushu Electric Industry to correct the present conditions, by application of the Waste Disposal and Public Cleansing Law, but Kumamoto City found it difficult to enforce the act because, during the process of the investigation, they were not able to confirm the dumping of trichloroethylene.[6] Later, Kyushu Electric Industry gave Kumamoto City ¥30 million toward the cost of groundwater conservation, but only as a contribution, which, again, means that Kyushu Electric Industry does not take formal responsibility for the pollution. One of the points at issue in the Water Pollution Control Law, even in the revised law, is confirmation that ". . . contaminated waste water has been infiltrating underground".

5. Groundwater Pollution Caused by a "High-Tech Industrial Park": Higashine in Yamagata Prefecture

In Tohoku District, as well as in Kyushu District, many plants related to the semiconductor industry have been built because of abundant manpower and abundant water resources. Near the interchange of an express highway and the airport, "high-tech industrial parks" stand in a row beside the road. The local government of Yamagata Prefecture, adjoining Yamagata Airport, developed the Ohmori Industrial Park so as to attract factories to the area and, since the inception of the industrial park in 1976, 16 companies, including high-tech industries, have established operations there. In November 1991, a local government investigation formally detected trichloroethylene in the groundwater of Higashine City for the first time. Yet, Yamagata Broadcasting Company had already reported on the TV program *Zoom in the Morning* on April 14, 1989, that 1,1,1-trichloroethane had been detected in the groundwater on the outskirts of Ohmori Industrial Park following its own investigation during March of that year. The local government investigation in 1992 reported that trichloroethylene (2 ppm maximum; equivalent to 67-fold the standard value) had been detected in the groundwater from wells at 35 places in an area 2.5 km east and west and 1 km north and south of the west side of Ohmori Industrial Park. During periodic monitoring investigations during 1992, ". . . the highest density of trichloroethylene had been detected"[7] and ". . . it was found that a wide area of Higashine was polluted".[8] At Ohmori Industrial Park, the suspected pollution source, five plants had been using organic solvents such as trichloroethylene. These plants were Yamagata Casio, which produces clocks, electronic notebooks, and metal molds and employs approximately 900 people, Higashine Shin Dengen, which produces semiconductors and employs approximately 320 people, Yamagata Sanken, which produces semiconductor chips and employs 550 people, Yamagata Fujitsu, which produces small-sized magnetic disks and employs approximately 1000 people, and Yamagata Kinseki, which produces crystal vibrators and crystal oscillators and employs approximately 500 people (data based on a telephone questionnaire conducted by Yamagata Broadcasting). The local government has already confirmed soil strata contamination at the sites of three of these business establishments. One of the enterprises dug up contaminated soil and pumped groundwater from August 1994, while another enterprise from August 1994 vacuumed up trichloroethylene that had seeped underground; both these operations have now finished. Another enterprise has also started to dig up contaminated soil and allow the release of trichloroethylene gas. This area used to be richly abundant in groundwater resources so good in quality that its water was suitable for *fu*, a special product of the district. Of the 35 wells in which the concentration of trichloroethylene exceeded the standard value, public tap water has been substituted for well water (on which the plants depended entirely) in nine cases, at a cost to the city (according to the Living Environment Section of Higashine City). Some neighboring business establishments still utilize contaminated well water, but not for drinking, and Ohmori Industrial Park, despite its being the pollution source, continues to utilize as much as 4500 tons of groundwater per day (Fig. 5.3).

The Environmental Protection Agreement between Higashine City and Ohmori Industrial Park has had little effect on the groundwater pollution problem, and, as a

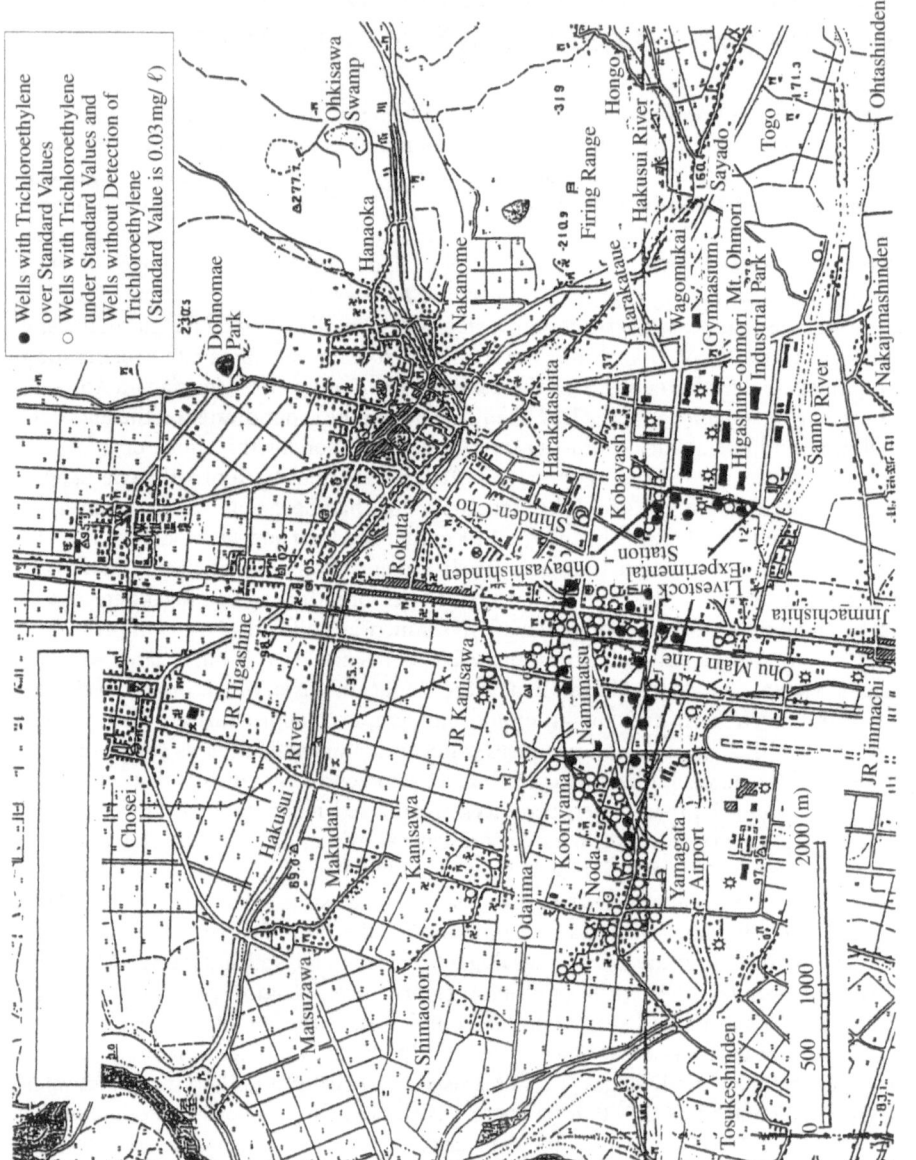

Fɪɢ. 5.3. Groundwater pollution at Higashine (Yamagata Prefecture, 1993)

result, the Yamagata Local Government is solely in charge of investigation and guidance. The local government, however, has not yet made public how far the correction of the pollution source and the pollution mechanism has proceeded. As a consequence of the yen's recent appreciation and "Industrial Cavitation", industrial overseas transfer has been promoted, so that some related plants in the Tohoku Electronic Zone have been forced to cut down on domestic production and people in the area cannot help but get nervous about the reduction in size or closure of the plants. That is why both the local government and the residents have taken an "indecisive attitude" toward the plants.

6. Groundwater Pollution Caused by Mitsubishi's "High-Tech Research Development Plant"

Whereas the semiconductor industry has undergone global development and many Japanese industries have recently been established in Southeast Asia, domestic semiconductor plants are making desperate efforts to survive as "Research Development Base Plants". For instance, Mitsubishi Electric Kita-itami Plant, located in Izuhara of Itami, Hyogo Prefecture, which employs 3000 people, is not only the base of research, development, and trial production related to the semiconductor section of Mitsubishi Electric Company, but is also a continuous mass production base of chemical compounds and hybrid semiconductors.

Geo-pollution caused by trichloroethylene has also occurred at the Kita-itami Plant. Groundwater pollution in Itami has been public knowledge ever since Konyo, Higashikuwano, Higashino, and Shimogawara in Itami were designated as areas for periodical groundwater monitoring investigation inquiries by the Hyogo Local Government. In 1989, 360 ppb trichloroethylene was detected in the groundwater at Higashino to the north of Mitsubishi Kita-itami Plant. An outline of the pollution was published, but the plant's name was withheld by the person in charge, who worked for the Hyogo Local Government.[9] The report said that, following an investigation of well water made by the Public Health Center in 1987, trichloroethylene was detected, after which time the Local Government Authority inspected all the plants near the contaminated well in order to locate the leak from the trichloroethylene rinsing tank at the Mitsubishi Kita-itami Plant that had occurred about 1980 and had already been secretly repaired. During a boring inspection within the plant, the investigators decided that the highest concentration of trichloroethylene in the soil was in the first aquifer. At the Kita-itami Plant, as much as 440 tons trichloroethylene had been used annually. By solidification of the contaminated soil, the concentration of trichloroethylene in the observed well has decreased, but pollution of the area around the well has not been halted. Nothing has been made public since the discovery of groundwater pollution caused by Mitsubishi Electric Kita-itami Plant. No detailed explanation of pollution has been given, but it can be said that groundwater around the well continues to be polluted, because 36 ppb trichloroethylene was detected in the groundwater from Higashino, to the north of Mitsubishi Electric Kita-itami Plant, in 1994.[10]

7. Groundwater Pollution of "A Chrysanthemum and High-Tech Town": Takefu, Fukui Prefecture

Takefu, a town of approximately 70000 people in Fukui Prefecture on the Japan Railways (JR) Hokuriku Line, a city famous for "Chrysanthemums and High Tech", is proud of its record as the prime shipper of manufactured products in Fukui Prefecture. Approximately 40% of the products shipped and the employees in the town are connected with the electric machine appliance industry. Following an investigation by the local government in 1989, groundwater pollution was discovered in Takefu: 85 ppb trichloroethylene, exceeding the standard value, was detected at Honbo, within the town's boundary.

At first, this was regarded as pollution by an unknown cause because no plants in Honbo used trichloroethylene. Professor Isamu Tsugo of Fukui Industrial Vocational High School, now Professor Emeritus, was asked to investigate the groundwater by the town Mayor, and he commenced a fundamental investigation in January 1990. Following an investigation of at least 200 wells, Professor Tsugo suspected that the pollution source was the Fukui Murata Plant, upstream of the groundwater, located at Okamoto of Takefu, producing ceramic condensers and employing 2600 people. Pollution had spread along the old river line of the Yoshinose River, between approximately 1 km at most and 100 m at least in width, and extending for approximately 6 km in length. The pollution had extended nearly as far as the tap water resources of the neighboring city, Sabae, and the highest detected concentration of trichloroethylene was 200 ppb[11] (Fig. 5.4).

When, under the guidance of Fukui Prefecture, the Fukui Murata Plant undertook boring investigations at 25 plots within its site, trichloroethylene was detected at all sites, in particular at three sites near the ceramic condenser plant, in the strata 2–14 m underground; the concentration of trichloroethylene ranged from 1000 to 2400 ppb. Contaminated soil spread over approximately 2000 m². The local government said that the ceramic condenser plant had stopped using trichloroethylene in 1986 and had remodeled the building as an office in 1988. The local government said that ". . . so as not to diffuse pollution by digging soil",[12] Murata undertook cleanup measures of vacuuming and pumping. Professor Tsugo's investigation revealed that the plant had been very careless about the way it had stored trichloroethylene, that it had scattered trichloroethylene as a herbicide, and that it had not paid the cost of changing from polluted groundwater to tap water. The 1993 groundwater periodical monitoring carried out by Fukui Prefecture said that although pollution as high as 81 ppb trichloroethylene had continued to spread for lack of fundamental prevention measures, owing to its sensitive economical dependence on Fukui Murata Plant, the Fukui Prefecture and Takefu City had taken an indecisive attitude toward groundwater pollution. Two more cases of large-scale groundwater pollution have been confirmed in Takefu. One is trichloroethylene groundwater pollution caused by Takefu Matsushita Electric Company, near the Oushio Station of the Hokuriku Line (located in Imajuku of Takefu, which produces power generators and electric motors, and employs approximately 1000 people). It is said that Takefu Matsushita Electric Company, concerned about the possible effects on the neighboring Hokuriku Coca-Cola Bottling Company, has already undertaken pumping measures to clean up the

FIG. 5.4. Groundwater pollution at Takefu City (Osamu Tsugo, 1990)

groundwater. Other cases of tetrachloroethylene groundwater pollution are those caused by the Takefu Plant of the Shin-etsu Chemical Industry, near JR Takefu Station (located in Kitafu of Takefu, which produces silicon and rare earth, and employs 500 people), and the Takefu Plant of Shin-etsu Semiconductor Industry.

Thus, Takefu, famous for "Chrysanthemums and High Tech," is urged to take fundamental measures against pollution if it hopes to avoid becoming known "the city of high-tech pollution".

8. The Most Serious Case of Japanese High-Tech Groundwater Pollution Over a Wide Area: the Outskirts of Yohkaichi, Shiga Prefecture

The outskirts of Yohkaichi, Shiga Prefecture, are the most widely polluted parts of Japan. According to the tap water source quality survey for all Japan, the concentration of trichloroethylene detected in the public water supply from southern sources at Azuchi, Shiga Prefecture is threefold the standard level. Upon further investigation of groundwater in private wells in Azuchi and on the outskirts of Yohkaichi and Ohmi-hachiman, concentrations of trichloroethylene over the standard level were detected in 27 wells over an area $40 \, km^2$ wide, and more than $10 \, km$ in length, from the outskirts of Yohkaichi Interchange of the Meishin Expressway to the City Area of Azuchi and the northwest parts of Ohmi-hachiman. The highest concentration of trichloroethylene was 1770 ppb within the city area of Yohkaichi, specifically Seiwa, whereas the most heavily polluted area was near Meishin Yohkaichi. Nine plants within the polluted area, located in the central industrial park, used trichloroethylene, and seven plants consumed 379 tons of the total amount of trichloroethylene used. Three of these seven electric machine appliance plants were: Yohkaichi Business Establishment of Murata Factory, located in Higashi-okino of Yohkaichi, producing ceramic condensers, resistors, and circuit products and employing 1300 people; Yohkaichi Factory of Kansai NEC Electric, located in Myohoji of Yohkaichi, producing diorde and employing 260 people; and Toppan Printing Electronics Plant, Myohoji of Yohkaichi, producing parts for electric appliances and communication machines, such as color filters, and employing more than 500 people (Fig. 5.5).

Of the remaining four plants, one is the Yohkaichi Factory of the Daishowa Paper Industry, producing pulp, paper, and paper-processed goods, located in Myohoji of Yohkaichi, and employing more than 500 people. Another is the Toyo Radiator Yohkaichi Plant, located in Gochi of Yohkaichi, producing transfer mechanical appliances, car coolers, and heat exchangers, and employing 500 people. A third is the Toppan Printing Shiga Plant, located in Myohoji Town of Yohkaichi, which is concerned with printing, publishing, and producing color filter. The fourth plant is Takiron Yohkaichi Plant, located in Higashi Okino of Yohkaichi, producing agricultural civil engineering material products, such as hard polyvinyl chloride (PVC) and coating iron wire, and employing 160 people (presumed from Table 4.2.3 in Japan Water Pollution Research Association[13]).

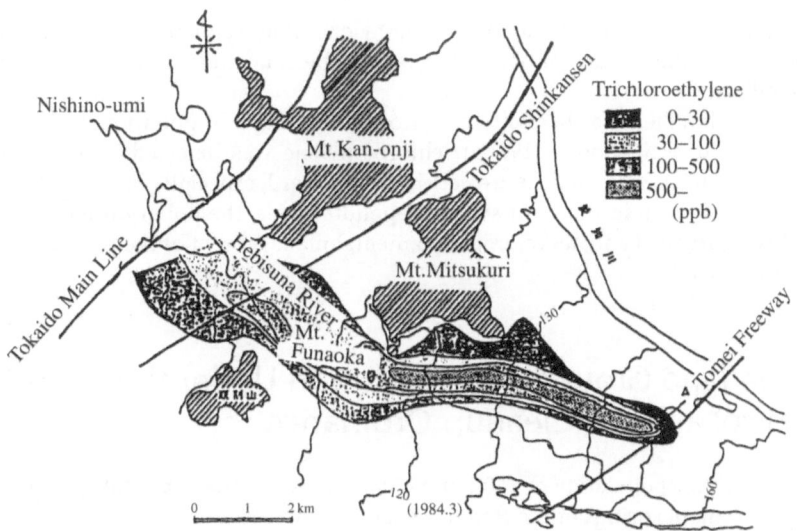

FIG. 5.5. Groundwater pollution near Yohkaichi by trichloroethylene (ppb) in 1984
Source: Japan Water Pollution Research Association (1986) Groundwater contamination measures research, case study. Tokyo, p 134

Two of the above-mentioned companies, namely Daishowa Paper Industry and Toyo Radiator, discharged water that was contaminated over the effluent standard, whereas two electric appliance plants and three companies, namely Daishowa Paper Industry, Toppan Printing, and Takiron, used a huge amount of groundwater exceeding the standard (from February to July 1984; presumed from Table 4.3.2 of Japan Water Pollution Research Association[14]). Finding it difficult to specify the cause of the pollution, Shiga Prefecture gave guidance on how to reduce or halt the seeping of organic solvent underground. Shiga Prefecture advised that water wells should be provided for two business establishments, that aerators should be newly equipped to four plants, that absorption of activated carbon should be installed in one plant, and that the depth of the water wells should be changed for three plants. By July 1985, no business that had trichloroethylene seeping underground exceeded the controlled target values, although, even at this time, we may notice that one business establishment still allowed organic solvent to seep underground. Subsequent investigation of the groundwater pollution revealed the following:

1. Trichloroethylene was detected at much higher concentrations than tetrachloroethylene and 1,1,1-trichloroethane, and it was distributed in a belt around the heavily polluted zone. The groundwater was polluted, without any diffusion, in the direction of the upper to the lower reaches of the old river basin.

2. There was little change in the distribution of contamination by trichloroethylene. These findings may be because, in the aquifer of the soil near the pol-

lution sources, the insoluble trichloroethylene that remained began to dissolve gradually, owing to fluctuations of the water table, and was carried by the stream of groundwater.

Even today, there has been no reduction in the scope of pollution.[15] In the Shiga Prefecture investigation of 1994, trichloroethylene was detected in 15 of 20 areas examined and, to our surprise, exceeded the standard, especially in six of the 15 areas examined.[16] Even in the most seriously polluted area, the pollution mechanism has not been corrected and, therefore, fundamental management of the problem has been left unchanged.

9. Hadano Groundwater Pollution Under the Pollution Control and the Cleanup Ordinance

Hadano, with a population of 160 000, a city famous for "the beautiful spring of Kobo" and "a group of springs of the Hadano Basin", is located at the foot of Tanzawa in the mountainous district of Kanagawa. In January 1989, a weekly photography magazine made public the news that Hadano spring was polluted by tetrachloroethylene (34 ppb). Yet, the 1983 Japan tap water source quality survey had already confirmed that Hadano was polluted by organochlorine chemical substances, although the level of contamination was below the prescribed standard.

Since Hadano had set up a water supply business in 1890, 65% of its tap water source has been supplied by groundwater from the "natural water jar". The city is equipped with 56 fountain head wells, 29 water distribution systems, and many private wells. Since groundwater pollution was confirmed in 1989, Hadano has installed aerators to four of its water distribution systems and transferred private wells to the water supply system. The industrial park, where many business establishments have continued to use chlorinated chemical substances, is located in a groundwater recharge area at the center of the basin, and the polluted area consists of approximately 12 km^2 within the city along both sides of the Mizunashi River running through the city center. Hadano, which is taking the situation very seriously, immediately investigated the matter in order to combat the groundwater pollution mechanism. Since 1990, Hadano City has continued to analyze the area, by boring and installing observation wells.[17]

An investigation of the groundwater along the left bank of the Mizunashi River running through Soyahara Industrial Park reported the following:

1. That, in spite of the very narrow pathway of the second gravel layer, pollution infiltrated the third gravel layer vertically in times of drought, every 6 months;
2. The river flows slowly through the third gravel layer, to which groundwater pollution has spread; and
3. The third gravel layer was polluted by the upstream waters of the Mizunashi River along the 14th Municipal Road, and the scope for pollution of the third gravel layer has spread rather widely along the left bank.

The right bank of the Mizunashi River, where Hadano Technical Park and Hirasawa Industrial Park are located, was found to be polluted:

1. The pollution pathway and scope for pollution were very narrow in both the second and third gravel layers; and
2. The areas of Hirasawa and Imaizumi spring water were assumed to be contaminated by pollution of the right bank and pollution of the fourth gravel layer of the left bank.

The Japan Water Environment Association confirmed, in 1992, that 60-m deep wells in the left bank of Mizunashi River were polluted by 95–866 ppb trichloroethylene and 129–143 ppb tetrachloroethylene. This report[18] established the depth and the area of pollution.

Hadano inspected those business establishments that were using chlorinated compounds and, consequently, by 1992, 58 business establishments were forced to stop using chlorinated compounds, 27 substituted another substance, 31 improved their methods of use, and 27 improved their methods of storage. At the same time, Hadano confirmed that 63 business establishments had carried out simple surface soil investigations with a boring bar detection tube, compared with 131 business establishments in total that were still using or had used chlorinated compounds, in order to make "a general investigation" of groundwater pollution, and that the work was covered by the municipal budget. In addition, Hadano City inspected storage of new and waste liquid that was then either in use or out of use. The types of business and business establishments were classified, with some exceptions, into three main types: metal product manufacturing, electric mechanical appliance manufacturing, and transport mechanical appliance manufacturing (according to the Environmental Conservation Section of the Environment Department of Hadano). Hadano undertook a fundamental investigation of 63 companies (the cost was covered by the City), and this included boring (from 1991 to 1994). Forty-four of the 63 companies made a "detailed investigation" by themselves, at their own cost, and consequently started a cleanup business of their own (the pollution at the remaining companies was cleaned up by the Fund for groundwater contamination control cleanup).

The methods used by the 44 companies who, by 1995, had enforced a cleanup business, include low heat processing, adopted by three companies, industrial waste treatment, adopted by five companies, soil containment gas absorption processing, adopted by two companies, original location gas absorption processing, adopted by 33 companies, and pumping aeration processing, adopted by one company.[19] After 22 companies, including two laundries, had completed the cleanup, approximately 9 tons organic solvent had been collected.

In Hadano Industrial Park, regarded as the groundwater pollution source, many major high-tech manufacturers and related companies have been built. On the left bank of the Mizunashi River, we can find, among others, Topra (manufacturing small screws and bolts, and employing 230 people), Shinko Air Conditioning Industry (manufacturing air-conditioning appliances), Stanley Electric (manufacturing car electric products and semiconductors, and employing 1700 people), Nikko Electric Industry (manufacturing internal combustion engine electric products, and employing approximately 200 people), Yokokawa Electric Appliance (manufacturing defense appliances, and employing approximately 30 people), Nittan Valve (manufacturing valves), Toshiba Ceramic (manufacturing quartz glass, and employing 140 people), Toyo Radiator (manufacturing radiators and oil coolers, and employing 500 people), and Nippon

Inter (manufacturing semiconductor commutators, and employing 640 people). Still in Hadano Techno Park, on the right bank of the Mizunashi River, we find Shimazu Plant (manufacturing electromagnetic analysis equipment and semiconductor manufacture equipment, and employing 200 people), Nissan Car Body (manufacturing car parts, and employing 80 people), Tokyo Electric (manufacturing cleaners and small motors, and employing 600 people), and Hitachi Plant (an electronic computer system manufacturer). In Hirasawa Industrial Park, Kobe Steel has built a plant for the manufacture of steel pipes for air-conditioning systems that employs 430 people. The names of the business establishments that have dealt with the cleanup of groundwater pollution have not been made public, but no doubt some of the above-mentioned enterprises will be among them.

Hadano, which has already established a Groundwater Usage Cooperation Fund System for groundwater preservation and has taken the opportunity to check groundwater pollution, has led the country in enforcing the Groundwater Contamination Control Cleanup Ordinance, the "Japanese Superfund". From 1989 to 1995, Hadano spent approximately ¥360 million in total for the investigation of groundwater pollution, and has saved approximately ¥50 million of the fund. The Hadano Groundwater Pollution Cleanup Business has been run on one-tenth of the expenditure estimated in 1991. This is partly because a very efficient cleanup system has been developed by the enterprises that have operated at an unexpectedly low cost. Nevertheless, problems regarding the solution for the mechanisms of pollution, cleanup measures, and the question of public information remain. Taking no account of its city's significant dependence on groundwater, the Hadano Local Government developed a high-tech Industrial Park just within the groundwater recharge area, which must be the principle cause of the complicated pollution detected. When we consider the current general use of organochlorine chemical compounds, if the same investigation as that performed by Hadano is carried out in other areas, geo-pollution is sure to be found everywhere.

10. Groundwater Pollution by High-Tech Subcontractors

Because high-tech products are made up of many individual parts, many subcontractors have built factories all over Japan and throughout Asia. This means that groundwater pollution is likely to be caused by high-tech subcontractors not only in big cities, but also throughout Japan. For instance, in Fukushima Prefecture, the Fukushima Local Government required enterprises to pay the cost for the cleanup of the strata and groundwater polluted by organochlorine chemical compounds. The enterprises affected by the enforcement of the cleanup extended to 38 districts.[20] Three of the cases related to the makers of electronic parts. One of the manufacturing plants, in Itozawa, Tajima, South Aizu County, deals with communication and electronic appliances. After the discovery of pollution in 1989, the plant cleaned up the polluted strata. Another of the plants, at Kinosaki, Naganuma, Iwase County, manufactures computer terminal units. When groundwater was found to be polluted by trichloroethylene, this plant pumped out the polluted groundwater.[21] The Bullet Train Motor Manufacturing Plant of Hattanda, Kawamata, Date County, with approximately 200 employees, has an association with Toshiba. In 1989, the plant was found to be polluted by tetrachloroethylene. In addition, in Fukushima Prefecture, much groundwater pollution has been caused by subcontractors of precision machinery related to the manufacture of lenses.

In Machida, Tokyo, groundwater pollution has spread around the areas of Tsuruma and Kanamori.[22] The pollution mechanism was investigated by the Environmental Science Institute of Tokyo, which plans to solve the groundwater pollution mechanism relating to trichloroethylene, first detected at the site of the electronic part plant 10 years ago. The Institute has reported that, in Machida City, the combined effect of the amount of precipitation, the groundwater level, and the concentration of pollutants following the rise of the groundwater table reacted on the stagnant silty part of water permeability, which resulted in an increase in the concentration of the pollutant in the groundwater. In Mobara City, Chiba Prefecture, upon investigation of the pollution mechanism and upon investigation of the distribution of ground pollution in the strata pollution plume and in surface soil pollution, the Electronic Appliance Parts Manufacturing Plant, which uses trichloroethylene (Toshiba Component Cooperative Plant, employing 40 people), was named as the source of the pollution.[23]

Similarly, in Awa-chikura, Chiba Prefecture, the groundwater pollution mechanism caused by trichloroethylene was corrected, and the Condenser Plant (an affiliated company of Hitachi AIC, employing 60 people) carried out the cleanup by digging up and removing contaminated strata, by ground air absorption, and by using a suction well for drawing off contaminated groundwater.

As to the groundwater of Tochigi City and Tsuga Town, Tochigi Prefecture, both partially contaminated by trichloroethylene, considerable trichloroethylene was detected at concentrations as high as 4400 ppb in a drain from a side ditch at the Electrolysis Condenser Company, Nishikata, Kamitsuga County, after which a large amount of trichloroethylene was also detected in the soil strata. The plant granted the residents an on-the-spot inspection at the request of the Tomihari neighborhood self-governing body of Tsuga Town and presented an improvement plan to the Tochigi Prefecture. The improvement plan says that, in order to prevent underground leaking, the company installed a stainless steel saucer under the trichloroethylene rinsing tank and a tied storage tank with a pipe lest the leaking trichloroethylene should pollute groundwater. The drain, which carried highly condensed trichloroethylene and contaminated the soil strata, was pumped dry and aerated.[24]

In March 1992, in Fukushima, Yatsuo Town, Toyama Prefecture, the groundwater of an electronic parts plant was found to be contaminated by 43 ppb trichloroethylene. Trichloroethylene was detected in samples of 12 soil strata from 19 locales within the factory site, and the concentration of trichloroethylene was highest on the outskirts of the trichloroethylene depository (gas concentration 300 ppm). Forty tons of contaminated soil strata was removed and a follow-up survey of the contaminated well was maintained; trichloroethylene, at concentrations ranging from 35 to 92 ppb, was detected in 1993.[25,26]

11. Groundwater Pollution by Lens Plants

In July 1990, tetrachloroethylene was detected in the groundwater at Shirakuwata and the Sendo district of Kanuma, Tochigi Prefecture, whose specialty product, Kanuma Soil, is well known. After the Housing Reclamation Trader (Toda Construction) asked the Public Health Center for a groundwater quality test, the groundwater turned out to be polluted (7850 ppb highest concentration of tetrachloroethylene). Pollution had spread from the area of the intersection of the Tohoku Driveway and the JR Nikko Line to the low land of the Takeshi River Valley, an area approximately 2 km long and 300 m wide. Because well water was used by 56 of 186 households in this district, the residents panicked as soon as they knew of the groundwater pollution.[27,28] Although tetrachloroethylene is generally known as a contaminant of the cleaning industry, in the case of Kanuma, the investigation undertaken by Tochigi Prefecture assumed that the Canon Kanuma Plant, upstream of the groundwater and, at that time, employing 220 people, was the pollution source. Following a detailed investigation, the ground-water of 21 wells supplying 369 households in the area was found to have concentrations of tetrachloroethylene that exceeded the standard value. The Canon Kanuma Plant consumed 240 tons tetrachloroethylene annually in an undiluted solution for cleansing the grinding process of copy machines and camera lenses. The Canon Kanuma Plant absorbed the tetrachloroethylene by activated carbon and then poured the treated wastewater into a side ditch. Tetrachloroethylene was detected in the waste-water at concentrations exceeding the standard value. The undiluted solution, which was used for cleansing, was held in six underground waste liquid tanks, which were collected by an industrial waste disposal business. There was, however, a leak from both the cleaner waste liquid tank and the distillation reprocessing apparatus, which were then under repair. Under the improvement order passed by Tochigi Prefecture, the plant presented an improvement plan to set up a new aerator at the final discharge outlet, to exchange activated carbon for fresh carbon, and to set up an aerator in the coagulating sedimentation disposal process. Although the plant was closed the following year, pumping aeration disposal was started from January 1991, by digging three barrier wells within the plant site as a means to abate groundwater pollution. At the same time, pumping aeration disposal was carried out to set up a new barrier well outside the plant. The residents established the Tsuda-Sendo District Pollution Countermeasures Committee and negotiated with Canon through Kanuma City Hall. As a result of this negotiation, Canon prepared a memorandum in September 1990 to the effect that Canon would pay ¥600000 per household, ¥28 million in total, in compensation for pollution damage. As for health damage, 70 people in the polluted area underwent medical examinations at the end of 1990 and no abnormal symptoms were found.[29]

However, the pollution mechanism solution has still not been made public and, therefore, the extent of groundwater pollution is not clear; as a result, Tochigi Prefecture has not been able to designate Canon as the pollution source. Now that damage compensation has been paid, the cleanup has been deferred. The cost of long-term groundwater aeration pumping is likely to be covered by public expenditure. The case of Canon Kanuma Plant is characteristic of leaks from underground tanks storing tetrachloroethylene used for lens cleaning.

A similar case of groundwater pollution occurred at the Canon Fukushima Plant, with approximately 1500 employees. The plant used to produce a single lens reflex camera and an 8-mm video camera, but has now converted to the exclusive production of 8-mm video cameras. An investigation by Professor N. Chuman, the Administration and Sociology Department of Fukushima University, found groundwater pollution by trichloroethylene and 1,1,1-trichloroethane near the plant at Sakurashita,

Fukushima City, covering an area approximately 1.3 km long, approximately 100–400 m, and approximately 35 ha in total.[30]

As the Canon Fukushima Plant did admit to being responsible for the pollution, they have covered the cost of rechanneling the water supply in the polluted area. In Fukushima Prefecture, much groundwater pollution has been caused by small- and medium-scale lens makers in addition to the above-mentioned case. The situation in Fukushima Prefecture is not exceptional from a nation-wide point of view, and it is thought that extensive groundwater pollution caused by small- and medium-sized subcontractors related to the production of precision machines will have spread throughout Japan.

12. Conclusion

As we can see from the above account, pollution caused by high-tech plants and related subcontracting plants is certain to have spread throughout Japan, although, apart from the cases mentioned here, most instances of pollution have not yet been made public. As I argued in Chapter 4 we need to combat and correct the pollution mechanisms in terms of their importance in causing geo-pollution and the significance of the cleanup, we must put the corresponding cleanup countermeasures into effect, and we must establish a responsible guidance and regulatory organization system that can convince residents of the effectiveness of the treatment. Among the prefectures, there are differences in the degree to which solutions devised and pollution mechanisms are made public. This is due to the different attitudes of authorities toward the polluter, the various degrees of public dependence on groundwater, and the legal provisions fixed by such edicts as the Groundwater Pollution Control Ordinance.

It would be a sizable future assignment to proceed with groundwater cleanup according to the partially amended Water Pollution Control Law. As for the PPP, some enterprises refuse to admit their responsibility for pollution officially, and instead pay some of the cleanup expenses as a "contribution". It would be much more important for the administration to clarify the relation between pollution and its past causal beginnings. In the case of small- and medium-scale enterprises that are unable to clean up the pollution, the responsibility of the related parent company or the financing company should be the subject of investigation, as it is in the US.

Although some local government bodies have raised a fund and established a system of renting clean-up facilities, the fund and the system are based on public funds, and it is indispensable that the responsibility of the enterprise as polluter is made clear and not to hide the information about pollution mechanisms and purification measures from the public.

References (all in Japanese)

1. Department of Environment, Kimitsu City (1993) The first report of cleanup measures of geo-pollution. Kimitsu
2. Japan Water Pollution Research Association (1990) Groundwater safety research, case study. A report to Environment Agency, Tokyo, pp 67, 102, 107

3. Sakura City (1995) White paper of environment, No 17. Sakura
4. Watanabe S et al. (1994) Cleanup of geo-pollution by chlorinated solvents Kumamoto. *Kikan Kankyo Kenkyu* [Q Environ Study], 95:53–69
5. (November 1, 1991) Cleanup experiment will start later this month. Kumamoto Nichi-nichi Shimbun
6. (December 13, 1991) How about companies responsibility? Kumamoto Nichi-nichi Shimbun
7. Yamagata Prefecture (1996) Survey of groundwater quality. Yamagata
8. Yamagata Prefecture (1995) White paper of environment. Yamagata, pp 33–34
9. Kobayashi T (1993) Restoration of groundwater contamination by trichloroethylene: situation and cleanup measures of groundwater and soil contamination. Kogyo Gijutsukai, Tokyo, pp 402–405
10. Hyogo Prefecture (1995) White paper of environment. Kobe, p 53
11. Okumura M, Tsugo I (1992) Survey for groundwater pollution by trichloroethylene in Tan-nan area. The annual study report of Fukui Industrial Vocational High School, No 26. Sabae, pp 199–208
12. (July 7, 1993) Chlorinated solvents is detected at factory sites. Fukui Shimbun
13. Japan Water Pollution Research Association (1986) Groundwater contamination measures research, case study. Tokyo, pp 116–142
14. Ibid.
15. Yohkaichi City (1993) Environment of Yohkaichi. Yohkaichi, p 18
16. Shiga Prefecture (1995) White paper of environment. Ohtsu, p 112
17. Hadano City (1993) General view of environment countermeasures. Hadano, p 78
18. Japan Water Pollution Research Association (1992) Soil and groundwater research, case study. p 29
19. Hadano City (1996) Cleanup of geo-pollution by simple cleanup system. Hadano, p 87
20. (August 13, 1993) Rethinking of water, Fukushima City (3). Kahoku Shimpo
21. Fukushima Prefecture (1992) The annual report of environment center at Kohriyama, Fukushima Prefecture. Kohriyama, p 59
22. Department of Environment, Machida City (1990) Survey of groundwater contamination by chlorinated solvents. Machida
23. (December 8, 1989) Trichloroethylene has been detected at Mobara, pollution source is Mobara Giken. Asahi Shimbun, Chiba edition
24. (July 5, 1992) Groundwater contamination at Tsuga, Tochigi. Shimotsuke Shimbun
25. Toyama Prefecture (1992) The annual report of Toyama Environment Center, No 20. Toyama, pp 36–37
26. Toyama Prefecture (1993) The annual report of Toyama Environment Center, No 21. Toyama, p 35
27. (August 8, 1990) Groundwater contamination, fears spread. Shimotsuke Shimbun
28. (August 25, 1990) Canon is trying to improve facilities. Shimotsuke Shimbun
29. (January 25, 1991) Health effect is not confirmed. Shimotsuke Shimbun
30. Chuman N (1992) Groundwater contamination with chlorinated solvents in the southern Fukushima Basin, northern Japan. In: Fukushima University (ed.) The annual report of synthetic study at Fukushima University, nature and its humanization, No 3. Fukushima, pp 19–29

Chapter 6
Geo-Pollution and a Cleanup System*

1. Introduction: What Is Geo-Pollution?

In Japan, groundwater pollution has recently occurred on a nationwide scale.

Groundwater has been polluted by, for instance, chlorinated organic compounds such as trichloroethylene and nitrate. To counter this, a partial revision of the 1989 Water Pollution Control Law has implemented measures such as a ban on the infiltration underground of water containing hazardous substances.

Thus, the condition of the groundwater is now regulated by the law in the same way that a "public water body" has to be. Thanks to this regulation, all prefectures in Japan are now required to make public the annual measured values of groundwater quality. Subsequently, the serious occurrence of groundwater pollution in Japan has become more generally known.

To make matters worse, the pollution is not limited to groundwater. To begin with, groundwater pollutes the stratum with toxic substances and, then, as a matter of course, brings about ground air pollution. Consequently, the three instances of pollution in the groundwater, pollution in the sedimentary strata, and pollution in the ground air have been grouped together, and are now referred to as "geo-pollution". The effect of geo-pollution is to change the sedimentary strata (the cluster of soil particles), the groundwater, and the ground air into substances that are physically, chemically, and biologically disadvantageous to humans, as well as inducing a chemical reaction between artificial substances.

It can then be deduced that a special type of soil pollution occurs in the sedimentary strata in which various chemical substances in the soil are transformed. As polluted substances in the water and ground air run through stratum gaps and stick to stratum components, they work to change the quality of those (strata) components, and result in the harmful phenomenon now known as geo-pollution. In other words, groundwater pollution occurs when polluted substances and water are dissolved or mixed with interstitial water in the sedimentary strata.

Sedimentary strata pollution by chlorinated organic compounds occurs when the sedimentary strata, as the first pollution source, works as the secondary polluter of groundwater.[1]

---

* Originally published in *Economic Journal of Hokkaido University* (Sapporo, 1996) 25:61–82. With the permission of the University.

Meanwhile, geo-pollution can be correlated with surface water during the process of circulation between land and water. The incessant fluctuation of atmospheric pressure leads to the inflow and outflow of polluted ground air, and the atmosphere is consequently polluted. In the end, polluted substances seep up to the surface of the earth; we call this "cross-media pollution".[2,3]

The investigation carried out in this chapter has three purposes: (1) to indicate the present situation of geo-pollution in Japan and the significant need for a cleanup of geo-pollution; (2) to scrutinize the abatement policies and the points at issue of geo-pollution as instituted or recognized by nationwide and local governments; and (3) to elucidate those problems caused by geo-pollution that need to be solved with dispatch.

2. Spread and Kinds of Geo-Pollution

2.1 The Present Situation

According to the findings of a survey on the general state of groundwater pollution carried out by the Environment Agency in 1994, concentrations of trichloroethylene in excess of water quality decreased from 0.9% in 1989 to 0.3% in 1993, whereas concentrations of tetrachloroethylene decreased from 1.2% to 0.5%, respectively.

It seems, at first glance, that the decreasing percentages revealed by the data signify a reduction in groundwater pollution. However, when we scrutinize the data, it becomes obvious that the areas investigated have been extended from the highly polluted urban areas originally specified to include the outskirts of towns (Table 6.1).

Because groundwater pollution caused by chlorinated organic compounds does not usually spread sideways, but penetrates vertically, the highly polluted area is limited. It is hard to detect pollution outside the periphery of the pollution source, unless it is within a designated speculative zone.[4]

TABLE 6.1. Waterworks polluted by trichloroethylene in Japan

Area	Drawing of drinking water from headwaters: practice discontinued	Treatment	Dilution	Monitoring	Total
Hokkaido					
Tohoku				1	1
Kanto	14	21	1	14	50
Chubu	7	7	1	5	20
Kinki	4	8		3	15
Chugoku	1				1
Shikoku				1	1
Kyushu	1	3			4
Okinawa					
Total	27	39	2	24	92

Source: Ministry of Health and Welfare, Waterworks Division, 1993
Note: The waterworks include simple waterworks and special waterworks
     The numbers show the waterworks where concentrations in excess of water quality standards were detected up to 1993

However, once a full-scale investigation of the urban area is made, organic solvents can be quite easily detected in the groundwater. An investigation of 108 wells in Kawasaki City, Kanagawa Prefecture, detected trichloroethylene in 52 wells (48%) and found that 20 wells (19%) fell short of the environmental standard quality level, whereas tetrachloroethylene was detected in 60 wells (56%), and 13 wells (12%) were found to fall short of the environmental standard.[5]

In my book *High-Tech Pollution*,[6] I blame geo-pollution mainly on the high-tech industry. Upon re-investigating the pollution problem, I can report that "high-tech pollution" has become more serious without our having realized it.

Yet, there is also no doubt that further investigation is likely to reveal many pollution sources around us quite apart from the high-tech industry.

As for high-tech pollution itself, in addition to what is known of Toshiba, Taishi Semiconductor Plant of Taishi in Hyogo Prefecture[7] and Toshiba Components of Kimitsu, Chiba Prefecture,[8] which led to the partial revision of Water Pollution Control Law, recent research in Kumamoto has partially uncovered the mechanism of groundwater pollution.[9]

It is well known that geo-pollution and groundwater pollution are caused by semiconductor and electric parts plants all over the country: Yohkaichi in Shiga Prefecture,[10] Takefu in Fukui Prefecture,[11] Higashine in Yamagata Prefecture,[12] Hadano in Kanagawa Prefecture,[13] and Itami in Hyogo Prefecture,[14] to name but a few.

Many related pollution problems also occur at medium- and small-sized plants, the so-called high-tech subcontract plants, such as Machida in Tokyo,[15] Kumamoto area, Hadano in Fukushima Prefecture, Ichikawa in Hyogo Prefecture,[16] Mobara in Chiba Prefecture,[17] Nishikata in Tochigi Prefecture,[18] Tokura in Nagano Prefecture,[19] Yatsuo in Toyama Prefecture,[20] and Shuto in Yamaguchi Prefecture.[21]

The most common cause of groundwater pollution is the leaching of tetrachloroethylene, the detergent used by the cleaning industry. Because most laundries are medium- or small-sized businesses with few funds, they cannot afford to grapple with the cleanup or help to illuminate the pollution mechanism, and because very little well water is used in urban areas, outbreaks of geo-pollution are left untreated.

Pollution by tetrachloroethylene also occurs in the textile industry: for example, at Tsukidate in Miyagi Prefecture,[22] Yonezawa in Yamagata Prefecture,[23] Tohkamachi and Gosen in Niigata,[24] and Kyoto.[25]

Because chlorinated organic solvents have been used widely and stored very carelessly, even in the metal industries, general machinery, and precision machine companies, pollution is apt to occur in small plants such as city ironworks no less than in major plants. A well-known case of pollution is that of the metal hardware manufacturing factories at Tsubame in Niigata Prefecture.[26]

In addition, many pollution problems have arisen at the automobile parts plants at Odawara in Kanagawa Prefecture,[27] Nishiki in Kumamoto Prefecture,[28] Narashino in Chiba Prefecture,[29] and at Ohno in Fukui Prefecture.

An inquiry into the siting of an automobile parts plant was brought to court because of the problem of possible groundwater pollution.[30]

To date, we can locate the areas in Japan where pollution occurs and we are able to identify the probable causes of the pollution as follows:

1. Metal processing plants: Naganuma in Chiba City,[31] Samukawa in Kanagawa Prefecture,[32] Fujieda in Shizuoka Prefecture,[33] Fujinomiya in Shizuoka Prefecture,[34] Kyoto,[35] and Miki in Hyogo Prefecture;[36]
2. Surface treatment plants: Yokohama,[37] Nagano,[38] and Fujieda in Shizuoka Prefecture;[39]
3. Electric wire plants: Gotenba in Shizuoka Prefecture,[40] Itami in Hyogo Prefecture,[41] Tsuchiura in Ibaragi Prefecture,[42] and Tabuse in Yamaguchi Prefecture;[43]
4. Lens plants: Kanuma in Tochigi Prefecture[44] and Fukushima in Fukushima Prefecture;[45]
5. Watch-making plants: Fukushima Prefecture, and Kimitsu in Chiba Prefecture;[46]
6. Machine tool plants: Narashino in Chiba Prefecture;[47] and
7. Painting plants: Tsuru in Yamanashi.[48]

Unexpectedly, the chemical industry has also been pointed out as another source of industrial pollution. To our surprise, pharmaceutical plants, such as those at Takatsuki in Osaka Prefecture[49] and Amagi in Fukuoka Prefecture,[50] and other chemical plants, like those at Kashiwa in Chiba Prefecture,[51] have also been named. In addition, pollution is caused by general chemical product manufacturers, such as Kohriyama in Fukushima Prefecture, Himeji in Hyogo Prefecture,[52] and Fujikawa in Shizuoka Prefecture.[53]

Apart from what an ordinal industrial classification reveals, pollution has been found in the recycling of organic solvent itself, for instance, at Matsudo in Chiba Prefecture,[54] Kawasaki in Kanagawa Prefecture,[55] and Yamashina in Kyoto.[56]

The most important pollution sources, apart from industrial pollution, are landfill sites for industrial waste, and, at Ohmachi in Nagano Prefecture, even the headwaters are polluted.[57]

According to an *Investigation Survey of Appropriate Management Measures for Landfill Sites*[58] of the EX Metropolitan Institute, there have been many occasions when final landfill sites have caused pollution of both groundwater and surface running water: (1) polluted water leaked from a broken tank in a seeping water treatment plant; (2) many fireflies raised by firefly breeders died from polluted water; (3) a paddy field near a landfill site was damaged by salt from wastewater; (4) because of leaking water containing a high concentration of iron and manganese, one plant was forced to postpone the suspension of its operation; (5) while managing reclaimed land after completion of the reclamation, neighboring residents began to doubt the safety of the reclaimed land and finally came to the conclusion that the reclaimed land was, in fact, the source of the pollution in the neighboring groundwater; and (6) with the aim of establishing environment protection, seepage control work was introduced by covering reclaimed waste in a landfill site filled with non-flammable material with a polyethylene sheet.

Groundwater pollution occurred at a local site at Mizuho in Tokyo.[59] Similarly, at Hinode, in Tokyo, groundwater pollution has been caused by components other than chlorinated organic compounds produced at the landfill site.[60] At Tago in Katori, Chiba Prefecture, trichloroethylene and other chemicals have been detected in the water of wells in the neighborhood of the landfill site, and, as a result, local residents have asked for a shutdown of the operation.[61]

Other examples of illegally dumped waste as a source of pollution can be found at Iwaki in Fukushima Prefecture,[62] Kashiwa in Chiba Prefecture,[63] Kawanishi in Hyogo Prefecture,[64] and Karatsu in Saga Prefecture.[65]

A survey carried out by the Water Works Division of the Ministry of Health and Welfare reported that the number of water supply sources polluted by organic solvent amounted to 92 for March 1993, and, in detail, 50 were in the Kanto District, 20 were in the Central District, and 15 were in the Kinki District (see Table 6.1).

One more important pollution source that we cannot disregard is the military base. It cannot be denied that pollution in a military base is connected with the use of heavy metals, polychlorinated biphenyl (PCB), and asbestos, quite apart from organic compounds. It is assumed that the American military supply base at Sagamihara in Kanagawa Prefecture, for instance, is one of the sources of trichloroethylene pollution.[66]

A survey carried out by the Environment Agency reports that the types of businesses that cause geo-pollution are, among others, the chemical, electroplating, and electrical appliance manufacturing industries.

As pollutants, trichloroethylene and tetrachloroethylene have been newly added to heavy metal lead, hexavalent chromium (Cr^6), and quicksilver (see Table 6.2).

2.2 Harmfulness of Chlorinated Organic Compounds

Tetrachloroethylene, a chlorinated organic compound and a potential source of geo-pollution, has been utilized widely as a detergent by the cleaning industry.

Recently, there has been a reduction in both the quantity of output and the amount used, but the toxicity of tetrachloroethylene is greater than that of other chlorinated organic detergents.

The addition of approximately only one-tenth of a test tube (3 ml) of tetrachloroethylene to an ordinary 25-m long pool (approximately $460 m^3$ water) is enough to exceed environmental standards. According to an evaluation made by the US EPA, tetrachloroethylene belongs among the carcinogens in group B2, a classification that indicates sufficient evidence of carcinogenicity in animals, but inadequate or no evidence of carcinogenicity in humans.

According to an evaluation by the International Agency for Research on Cancer (IARC), tetrachloroethylene belongs to group 2A. This implies that tetrachloroetlylene is probably carcinogenic to humans.

Although trichloroethylene has been used widely as a detergent for washing semiconductors, electronic parts, and metal products, there has been a temporary reduction in both its production and consumption now that it is regulated as a designated chemical substance.

In addition, because 1,1,1-trichloroethane has been named recently by the Montreal Protocol as contributing to the greenhouse effect, the amount used has been reduced even more. Consequently, the amount of trichloroethylene used as a substitute for 1,1,1-trichloroethane is increasing.

Although trichloroethylene has been classified as a carcinogen for some time, according to the IARC evaluation, trichloroethylene belongs to group 2A (probably carcinogenic to humans).[67]

TABLE 6.2. Geo-pollution classified by industry and substance

Industry	Cases	Cd	CN	Pb	Cr^6	As	Hg	PCB	CCl_4	①	②	MC	TCE	PCE	Cu	Zn	Ni	Phenol	F	Oil	Other	Total
Textile	2	1	1	1		1	1	1						1				1				8
Wood process	2				2	2									1				1		1	7
Chemical	33	8	3	12	4	8	17	2				1	2	1	3	2	1	3		1	6	74
Oil, coal	1											1	1	1								3
Plastic	1	1			1																	2
Rubber	1												1	1								2
Ceramic	7	2		2	1	1	1					1	1	2								11
Steel	4	1	1	2	2	1	1									1	1					10
Nonferro-metal	9	4		5	1	4		1					2	1		2						20
Nonferro-mine	1	1																				1
Metal process	45	4	13	7	27	2	1	1				5	8	4		2						74
plating	(29)	(2)	(9)	(3)	(24)	(1)	(1)	(1)				(2)	(4)	(3)		(1)						(51)
General machine	2											1	1									2
Elect. machine	29	2	1	5	3		1	4				4	13	5		1						39
Transportation machine	8	1	1	2	3	2	1						2			1				1		14
Prec. machine	2											1	2	1								4
Gas	3		3				1													1		5
Recycle	4			1				2						1								4
Dry-cleaner	21											1	2	21								24
Waste disposal	8	2		2	1	2	1	1				1	2	2		1						15
Research institute	8			5			7	1					1				1			6	1	22
Others	12	3		2		1	2	3								3				1		15
Unknown	28	6		8	1	4	8	3	1	2	3	2	6	6	1	9		1		2		63
Total	232	36	23	54	46	28	42	19	1	2	3	18	44	47	5	22	3	5	1	12	8	419

Source: Environment Agency, 1995

Note: ①, 1,2-dichloroethane

②, *cis*-1,2-dichloroethane

TABLE 6.3. Production amounts of organic solvents used in Japan (tons)

Year	Trichloroethylene	Tetrachloroethylene	1,1,1-Trichloroethane	Dichloromethane
1990	56 850	83 619	184 991	77 466
1991	51 679	67 139	177 146	82 259
1992	61 080	63 225	168 440	83 519
1993	68 416	63 866	77 568	93 349
1994	77 159	57 777	54 629	88 877

Source: MITI, Statistical Year Book of Chemical Industry

It is evident that trichloroethylene is acutely toxic, and that even under standard permissible concentrations it inflicts various kinds of damage upon human health. In fact, many children at Woburn in the US have suffered from leukemia, caused by drinking well water contaminated with trichloroethylene. Consequently, it cannot be denied that there is an epidemiologically passive correlation between trichloroethylene and leukemia.[68]

In place of trichloroethylene and 1,1,1-trichloroethane, which have now been regulated, the use of dichloromethane has been rapidly increasing (see Table 6.3).

However, the problem still remains: it is not yet clear what effect dichloromethane has on the human central nervous system.

Thus, the regulation and use of organic solvents are trapped in what has been called a vicious circle, and the development of a non-organic solvent rinse without any chlorinated organic solvents is, therefore, badly needed.

New methods and techniques of rinsing are being gradually developed, including rinsing with alcohol or pure water, or manufacturing techniques without the need for rinsing.

2.3 The Significance of the Cleanup of Geo-Pollution

Once groundwater is proved to be polluted and is no longer suitable as drinking water, tap water can be easily substituted for groundwater. There is no doubt, however, that the source of the pollution is left untouched.

Even if it is easy and possible to switch the water supply from groundwater to tap water, tap water is sure to be contaminated with trihalomethanes, which lower the safety of the water and have a harmful effect on human health. Ignorance of geo-pollution and negligence of the appropriate cleanup of geo-pollution only accelerate the spread of the polluted areas.

As an example, at Fuchu, Tokyo, the authorities stopped drawing water from the well because the groundwater was contaminated with a chlorinated organic compound, but they subsequently neglected to take proper measures against the contaminated groundwater.[69] The result was the spread of contaminated groundwater throughout the city.

Because groundwater sources are connected with each other through underground veins, groundwater pollution spreads easily outside the administrative zone; for instance, at Yohkaichi, Ohmi-hachiman, and Azuchi in Shiga Prefecture,[70] Takefu and Sabae in Fukui Prefecture,[71] Narashino and Chiba in Chiba Prefecture,[72] Tochigi and

Nishikata in Tochigi Prefecture,[73] and Sagamihara and Zama in Kanagawa Prefecture.[74]

It is obvious from reports of investigations of sites polluted by groundwater, such as Kimitsu and Taishi,[75] that, below ground, trichloroethylene, a less-toxic chlorinated compound, is decomposed into 1,1-dichloroethylene, a more toxic chlorinated organic compound, which seems approximately 100-fold as carcinogenic as trichloroethylene.

This fact indicates that we have to consider the long-term effect of polluted groundwater on the environment, now that we are being confronted with the spread of polluted groundwater.

Yet, even when geo-pollution occurs around us, many people refuse to grapple either with the clarification of the pollution mechanism or with the cleanup of geo-pollution.

According to the *Report of an Administrative Inspection Result on Water Quality Protection Measures*,[76] the Administrative Inspection Bureau of the General Affairs Agency, which targeted 56 districts in 12 prefectures across Japan, found that local authorities offered three reasons as to why they did not investigate the groundwater pollution source: (1) it cost too much money to investigate (15 districts); (2) they did not know the procedure and the measures for specifying the pollution source (18 districts); and (3) water works are in common use in their district and so it is not always necessary to clean up the groundwater (10 districts).

Here, let me once more summarize the significance of the cleanup of geo-pollution.

The first reason is the issue of (the concern for) human health. Whether or not groundwater is required for drinking, groundwater pollution is linked to pollution of both sedimentary strata and ground air through the "geo-pollution of cross-media".

The second issue is how we should rank the importance in our lives of the precious quality of water. In Japan, the rate of groundwater supply for use in our everyday lives is approximately 30% and, in order to convert all drinking water into surface water, much more money is required. As the 1994 drought indicated, it is very difficult with the many geographical restrictions to tap much more surface water over a brief period of time. Groundwater is ranked highest of all sources of water because of its usually delicious, safe, and useful qualities.

The third issue, related to the second, is that the cleanup of geo-pollution can contribute to an elevation of the value of land and water resources. For instance, in the US, it is necessary to examine the state of land at the time of sale, and the consequent confirmation of any geo-pollution is indispensable if the sale is to go through. That is because the landowner is responsible for the cleanup of the land. It is a matter of course that once the land is polluted, the value of the property is diminished.

The fourth issue is that the chlorinated organic compounds that cause geo-pollution also contribute to global warming, as does 1,1,1-trichloroethane.

In order to preserve a clean environment globally, we should gradually cut down on the use of chlorinated organic compounds, clean up polluted soil, and recover toxic chemical compounds.[77]

3. Tackling the Cleanup

3.1 Problems at the National Level

One of the reasons why geo-pollution has been ignored in Japan, is that there are no regulations regarding geo-pollution at a national level.

Geo-pollution can be divided into two types: soil and groundwater pollution.

As far as soil pollution is concerned, in urban areas other than farmland, the Agricultural Land Soil Pollution Prevention Law is not applied, and there are no definite legal regulations for the cleanup of groundwater pollution.

Article 19 of the Water Pollution Control Law, revised in 1989, specifies the ". . . strict liability of a trader responsible for discharge and seepage into the underground to compensate for damages caused by discharge and seepage", yet the duty of cleanup is not specified.

The Waste Disposal and Public Cleansing Law provides, in Item 4 of Article 19 (Measure Order), that the government can order those responsible to take necessary measures to remove obstacles or prevent damage in the belief that the living environment would be or might be damaged.

This regulation, however, mentions only the disposal of present waste; this differs from the Superfund Law (CERCLA/SARA) of the US, and is not applicable to any pollution caused in the past.

Under these circumstances, geo-pollution in urban areas has been dealt with as a soil pollution problem only.

Until now, there have been no legal regulations regarding soil pollution in urban areas other than farmland.

In 1986, the Environment Agency announced Provisional Guidelines for Cleanup Targeted at the Contamination of State Property. In the guidelines, criteria for soil contamination and level of actions/measures for cleanup were provisionally proposed for nine contaminants, mainly heavy metals.

It was not until 1991 that an environment quality standard for soil pollution, required by the Basic Law of Environmental Pollution Control, was finally established. (This law names 10 substances, subsequently 25 substances, however, this standard is not applied to a landfill site.)

Along with this law, in 1992 the provisional guidelines of 1986 were revised and named Guidelines for the Cleanup of Soil Contamination of State Property (nine substances are named in these guidelines).

However, the provisional measure guidelines issued prior to the revision have been the principle guidelines used by most local governments for dealing with soil pollution of private land.

In November 1994, the Guidelines to Investigate and Measure for Soil and Groundwater Pollution were first issued. The main points of these guidelines are as follows:

1. Chlorinated organic compounds, as well as heavy metals, are subject to investigation;
2. With regard to chlorinated organic compounds, groundwater pollution and soil pollution are subject to investigation; and
3. Not only state property, but also land in general is subject to investigation.

Consequently, the earlier Guidelines for Cleanup of Soil Contamination of State Property were rescinded.

The new guidelines are a slight improvement upon previous guidelines because they mention "land in general", thus integrating soil and groundwater pollution with geo-pollution. However, they have neither a legal binding force nor do they mention the institutional issue of the person who is to be in charge of the cleanup and the allocation of costs.

The problems of comprehending the actual condition of groundwater pollution are as follows:

1. The Environment Agency will assist in bearing the cost of any groundwater quality investigation executed by the local government (the subsidiary rate is one-third of the total cost, and, in 1993, it amounted to approximately ¥90 million), and requires the Director of the Water Quality Preservation Bureau to inform it of the measured value of the water quality investigation and the locations of the wells by district; and

2. The Environment Agency is not really earnest in its attempt to grasp the extent of the pollution to be confirmed, the sources of pollution to be specified, and the cleanup action to be followed because no report of the investigation from the local government is required.[78]

The Environment Agency gives two plausible excuses for making slow progress in the cleanup of geo-pollution:[79]

1. The goal of cleanup has not yet been set; and
2. It has not yet been determined how the cost will be allocated.

This is how the Environment Agency explains the allocation of cost:

1. At present, it is difficult to specify the pollution source and the person responsible for the pollution;

2. It is hard to determine whether the pollution source is the result of illegal dumping, even when it can be specified as such, and it is still more difficult for the polluter to be specified;

3. General regulations for underground seepage control issued by the Water Pollution Control Law, and the waste disposal standards provided by the Waste Disposal and Public Cleansing Law, have already been published (there is, however, another consideration: whether it is reasonable to demand payment for the cost of cleanup for an offense committed prior to the drainage and the groundwater seepage regulation relating to toxic substances of trichloroethylene, and so on); and

4. It is very difficult to allocate cleanup costs for small-scale businesses.

If we assume that the cleanup will make little progress in the present serious situation because it has not been determined who will bear the cleanup cost, we have to admit that the doctor may well refuse to diagnose the patient's sickness until the patient's solvency is confirmed.

Although the real reason for the poor progress in cleanup is, therefore, the failure to recognize the importance of the pollution problem and the significance of its cleanup, the problem of allocating the cleanup cost needs, nevertheless, to be discussed separately.

Of the four points mentioned here, the first and second (the identification of the polluter) are both addressed with serious difficulty.

However, it is quite obvious from the example of Chiba that it is almost impossible to clarify the pollution mechanism without the concept of geo-pollution. Therefore, the question is similar to the problem of earlier illegal dumping: how to specify the unidentified polluter.

The third issue is the problem of retroaction, and the fourth is the problem of the solvency of small-scale businesses, both of which need to be discussed separately.

3.2 The Action of Local Governments: Technical and Financial Assistance to Clarify Pollution Mechanisms—The Chiba Method

The major obstacles to clarification of the geo-pollution mechanism are financial and technical problems. In order to solve these two problems, the Chiba Prefectural Government has established a technical and financial support system for its cities, towns, and villages, and, consequently, has achieved remarkable success in the clarification of both the pollution mechanism and cleanup.

The national government took the opportunity provided by the groundwater case brought against Toshiba Components in Kimitsu, Chiba Prefecture, to revise the Water Quality Pollution Control Law on a national scale. Revisions are bound to be partial— otherwise you would scrap the law and start again.

In Chiba Prefecture, the Guidelines for Groundwater Pollution Preventive Measures were enacted in 1989.

Based on these guidelines, various measures were taken by the Chiba Prefectural Government, such as monitoring groundwater quality, giving guidance to traders, measures to confirm pollution, and so on, as well giving technical and financial assistance to the cities, towns, and villages in Chiba Prefecture.

The details of technical assistances are: (1) to institute meetings for the study of groundwater pollution-preventive measures; (2) to assist in locating new sources of pollution; and (3) to investigate and clarify the pollution mechanisms and to offer guidelines for their remediation.

Of these three technical means of assistance, the third is the most essential.

The investigation teams are organized by the prefectural government, municipalities, and a private geological consultancy company in order to work out methods of investigation and enforcement of removal within the pollution site to be investigated. The team is a sort of "group of doctors", specializing in the cure of "geo-pollution". This type of technical assistance is possible because Chiba Prefectural Government has geo-environment specialists in its Water Quality Conservation Institute.

Simultaneously, Chiba Prefectural Government subsidizes the task of groundwater pollution prevention and assists in the physical examinations, which check the effects of trichloroethylene and other chemicals.

The previous subsidy assisted tasks for water quality investigation by confirming the actual nature of the pollution, and for the investigation of clarifying pollution mechanism (18 cities), as well as for measures to wipe out pollution (12 cities).

The rate of the subsidy has been proportional to the index of each financial power, ranging from 20% up to 70%, and it amounted to approximately ¥0.3 billion (54 cities) altogether as of the 1994 fiscal year.

It is obvious that, in spite of the small amount of the subsidy, the pollution mechanisms have been clarified in more than 20 cities and towns, and that in 12 cities and towns in particular, the cleanup of pollution is being promoted.[80]

3.3 Local Government Action: Japanese-type Superfund—Hadano City Ordinance

The most helpful reference for solving problems regarding the cleanup of geo-pollution in Japan is The Groundwater Pollution Prevention and Cleanup Ordinance promulgated by Hadano City, Kanagawa Prefecture. Hadano, which depends more on groundwater than any other district in Japan, has been afflicted by serious groundwater pollution. Hadano, herald to the whole nation, issued The Groundwater Pollution Prevention and Cleanup Ordinance in July 1993 (now, the Groundwater Conservation Ordinance).

The underlying principle of the ordinance is that the person whose actions in the past brought about geo-pollution has an obligation to make a detailed investigation and cleanup of the pollution (retroaction).

At the same time, the local government has established a fund made up of municipal contributions, while the Mayor, the Head of the Local Government, manages the issues of first making a detailed investigation of the situation and then cleaning up the polluted land, as well as allocating costs later when the polluter has been identified (Fund Method).

The purpose of the ordinance is, first of all, to ". . . prevent the pollution of groundwater caused by chemical compounds and to clean up pollution" (Article 1), which covers the whole problem of the geo-environment. Eleven compounds, including chlorinated organic compounds, are named as subject to regulation and investigation, as well as the companies that have used such materials (Article 2). In order to prevent geo-pollution, the ordinance stipulates that the founder of the plant has the obligation to report on the appropriate management of designated substances, as well as on the balance of the substances (Articles 4–20).

As to the investigation and cleanup of pollution, the ordinance requires the geo polluter to undertake a detailed investigation and any subsequent cleanup tasks as a result of previous, as well as present, actions (Article 22); this implies a retroactive system.

The Mayor is provided with the authority to make a fundamental investigation in order to grasp the general condition of the real pollution situation (Article 21). The trader is required to make a further detailed investigation and to run the cleanup business according to the "polluter pays principle" (PPP), while the Mayor is also required to endeavor to give whatever technical and financial assistance he can (Articles 24–26, Article 46).

The person deemed responsible for making the detailed investigation and running the cleanup business is the person (the trader concerned) who has been determined to be responsible for the geo-pollution of his land (the polluted land).

The action level is above the target level of cleanup recognized by the Mayor. Depending on the individual situation of the polluted land, the Mayor is also empowered to designate responsibility for the cleanup not only to businesses in the past or present who caused the pollution, but also to name anyone who polluted the geo-environment in the process of collection, transportation, and disposal of designated substances (Article 22).

Hence, ". . . the owner or possessor of polluted land has to cooperate with the related trader's or the Mayor's detailed investigation and cleanup business" (Article 35). This does not mean, however, that the owner or possessor of polluted land is charged with the direct responsibility of the cleanup. The designated trader is not able to make a plan and carry out and complete the cleanup business without the Mayor's approval and supervision (Articles 24–33).

Although a considerable number of factors have to be dealt with, such as the unidentified polluter, the unknown whereabouts of the related trader, the insolvent trader, and so on, the Mayor handles the business in their place.

However, when the polluter's whereabouts can be determined, the Mayor can allocate the costs to him later (Article 34). The cleanup process is undertaken when the composition of the groundwater in the soil and the strata is below the standard water quality of tap water (a list of Article enforcement regulations is attached to the ordinance). The level of cleanup of the soil, strata, and groundwater is provided numerically for each sampled test solution. In order to make sure that the business runs smoothly, the fund raised for groundwater pollution measures is financed by both the city and by the relevant traders' contribution.

The fund is allocated to the relevant traders' detailed investigation and cleanup business, as well as to the cost for investigation, cleanup, and health injury prevention supervised by the Mayor (Article 36–41).

The fund has planned up to ¥1 billion. The first proposal for a compulsory cooperative fund met with opposition from the Ministry of Home Affairs because it would entail a new local tax and, consequently, contributions were made optional.

As to "the announcement and the punishment", Article 51 provides that ". . . in the case of a geo-polluter the mayor can announce (make public) the situation through a public relations magazine, and that the person who neither makes a detailed investigation nor runs a cleanup business without a legitimate reason, is named an unscrupulous offender by the provisions of the Article". However, there is no mention whatsoever of any punishment.

This omission is based on the concept that the punishment of a polluter in the past violates the principle of "no punishment for retroaction"[81] (Constitution Articles 31, 39). At first, the concept of retroaction met with opposition from the Regional Legal Affairs Bureau, but was admitted on the condition that provision of punishment was excluded.

For all these reasons, Hadano City Ordinance is the most advanced in Japan from the viewpoint of investigation, cleanup, and funding.

In detail, the Hadano City Ordinance is notable for the following four points:

1. The detailed investigation and the cleanup business are, as a rule, established as the duty of the polluter (the trader or the manager of disposal), to which is attached the duty of retroactivity and joint liability;
2. In the case of a missing or insolvent polluter, the city runs the cleanup business;
3. To check pollution at the time of commencement and achieving the cleanup business, "the cleanup standard" has been established, based on the standard quality of tap water; and

4. A fund has been established to execute the business of the municipal and the polluter's cleanup.[82]

The Hadano City Ordinance has significance as a model for systematically solving any future geo-pollution problem.

4. The Problem to Be Solved

4.1 The Landowner's Cleanup Responsibility

What shall we do about the landowner's responsibility for cleanup when the landowner and the polluter are not the same person?

For the purpose of identifying the polluter, the US Superfund (CERCLA/SARA) holds the landowner responsible for cleanup; consequently, the responsibility for cleanup is very likely to be brought to court by the potentially responsible parties of polluter and landowner.

Dr. Hisashi Nirei of the Water Quality Conservation Institute of Chiba Prefecture sets out the five factors or conditions that require some sort of adjudication when deciding who should bear the cost of the cleanup in cases of geo-pollution:[83]

1. When pollution that has been caused by landowner's small enterprise lies outside the landowner's zone;
2. When the buyer of polluted land does not know that the land is polluted: in this case, as the US Superfund stipulates, the landowner is an "innocent purchaser";
3. When a landowner who did not pollute the land transfers it to someone else;
4. When a landowner who is also the polluter of the land sells the land and goes bankrupt; and
5. When illegal dumping pollutes a landowner's property.

In these cases, the landowners may have something in common.

In another Japanese instance, Yokohama City established Guidelines for Pollution Countermeasures at Landfill Sites (1986) in order to enforce entrepreneurs to check for soil pollution on quitting their land, and to take steps to abolish the plant.

In 1997, the number of entrepreneurs who had been obliged to check their holdings for soil pollution came to 134. 14 cases, mainly as a result of metal pollution, were under guidance, whereas 4 cases in which the soil had proved to be polluted were under investigation. Even in rare cases where landowners and entrepreneurs were not one and the same, all parties involved have cooperated in the investigation (according to the Water Quality and Geo-Environment Section of Yokohama Environment Conservation Bureau).

Kawasaki City authorities say in the Guidelines for Soil Pollution Countermeasures established in 1993 that the measures require the entrepreneur and the landowner to perform soil investigations and to dispose of polluted soil; the result of any transfer, abolition, or redevelopment of any factory must also be reported.

Also in Kawasaki, it does not matter whether the landowner is a different person from the entrepreneur.

However, the problems that arise over land transfer cases, having occurred once, are likely to occur more frequently from then on (according to the Water Quality Section of Pollution Dept. of Kawasaki Environment Conservation Bureau).

As to the landowner's responsibility for illegal dumping, according to the findings of the questionnaire on The Actual Situation and the Problem of Illegal Dumping sent out by the National Industrial Waste Federation, 60.9% (39 bodies all told) of all local governments must ". . . guide the landowner and the land manager in means of remediation" in the case of illegal dumping.[84]

Similarly, Professor Yoshinori Kitamura, who investigated the hearings of all national local governments, says that, as a result of administrative guidance, the landowners are actually required to treat pollution and mostly suffer the loss because all the cost falls on them.[85]

According to the explanation offered by Professor Tadashi Ohtsuka,[86] Article 717 of the Civil Law of Japan ("If any damage has been caused to another person by reason of any defect in the construction or maintenance of a structure on land, the person in possession of the structure shall be liable in compensation for damages to the injured party; however, if the person in possession has exercised due care in order to prevent the occurrence of such damage, compensation for the damage shall be made by the owner") and a claimable real rights (*dinglicher Anspruch* in German) of the existing law ("If, in the cases mentioned in the preceding two paragraphs, there exists any other person who is responsible for causing the damage, either the possessor or the owner may exercise the right to obtain reimbursement against such other person") impose both the liability of damages and/or the bearing of the cost of disturbance removal on the no-fault landowner.

However, these guidelines are limited only to cases where there is a fear of clear "damage" and "the danger of disturbance".

Professor Ohtsuka therefore suggests that the landowner should perform his supplementary duty only in the case of a missing, absent, and insolvent polluter, because it is fundamentally difficult to allocate the cost to the landowner who has contributed to pollution but without evidence of clear "damage" and the danger of disturbance.[87]

However, Professor Ohtsuka recognizes that if the cleanup cost is borne by the landowner, a sufficient pollution prevention incentive and a certain amount of cleanup incentive would need to be operating.[88]

I fundamentally agree with Professor Ohtsuka's view, in that it conforms to the PPP of imposing the most important responsibility on the polluter while it conforms to the actual Japanese situation and the Japanese law system, where the landowner has to perform his supplementary duty only in the case of a missing, absent, and insolvent polluter.

If it is possible that the landowner is required to do his duty of partially bearing the cleanup cost, the incentive both for pollution prevention and proper land management is surely raised, which is equivalent to the enactment of legislation to investigate soil pollution at the time of land sale.

The result would be an increase in the rate at which instances of soil pollution are found.

Trials similar to those brought on by the US Superfund are not always appropriate because they hold the landowner responsible for the cleanup, and are due, in large measure, to the particular features of American society and its law system.

4.2 Conclusion

The burden of this account has been to insist that the clue to geo-pollution is whether we are able to realize how important geo-pollution is and how significant the cleanup must be in consequence.

On the basis of a full recognition of the significance of what has been carried out in Chiba Prefecture, it is necessary to clarify the mechanisms that cause pollution, basing our account on the concept of geo-pollution.

In order to set about clarification, it is necessary to establish a Japanese-style Superfund system to conduct research and to clean up pollution, such as the one supported by a fund raised on the PPP as is required by the Hadano City Ordinance.

The establishment of an attainable Japanese-style Superfund system must be our present goal.

It is quite reasonable to expect that Japan, as a nation, should learn from these valuable experiences, and should establish a geo-pollution control and cleanup system as soon as possible, integrating in its program both groundwater pollution and soil pollution.

References (all in Japanese)

1. Nirei H et al. (1994) Groundwater contamination caused by VOCs. Env Res Qur 95:33–36
2. Suzuki Y (1993) Investigations and remediations at geo-pollution site. In: Geological Society of Japan (ed.) abstracts: the symposium on geo-pollution. Tokyo, p 58
3. Multi-media pollution is called multi-compartment pollution; see also: Urano K (1995) Multi-compartments pollution by volatile organic compounds. Waste Manage Res 6:13–23
4. Hirata T (1994) Research method of soil and groundwater pollution. Environ Measure Technol 21:39
5. Kawasaki City (1994) Findings of survey of groundwater pollution. Kawasaki
6. Yoshida F (1989) High-tech pollution. Iwanami Shoten, Tokyo
7. Japan Water Pollution Research Association (1990) The report to environment agency: research for groundwater pollution. Tokyo, pp 69–107
8. Kimitsu City (1993) Action report for groundwater cleanup. Kimitsu
9. Japan Water Environment Association (1992) The report to environment agency: research for hazardous chemical and groundwater pollution. Tokyo, pp 47–68
10. Japan Water Pollution Research Association (1988) The report to environment agency: research for groundwater protection. Tokyo, pp 116–142
11. Tsugo O et al. (1991) Groundwater pollution at the fan in Tannan, Fukui Prefecture. In: Proceedings of the 1st workshop on groundwater pollution and its prevention. Tokyo, pp 93–98
12. Yamagata Prefecture (1994) Findings of survey of groundwater pollution. Yamagata
13. Hadano City (1994) Summary of environmental pollution control. Hadano
14. Kobayashi M (1993) Remediation of groundwater pollution by trichloroethylene. In: The state and cleanup of groundwater pollution and soil pollution. Kogyogijutsukai, Tokyo, pp 398–409
15. Watanabe M (1992) A research on mechanism of groundwater pollution (2). Annual report of Tokyo Metropolitan Research Laboratory of Environment. Tokyo, pp 179–185

16. Ichihashi K et al. (1992) Situation of well pollution by trichloroethylene in Hyogo Prefecture. In: Proceedings of 2nd workshop on groundwater pollution and its prevention. pp 132–137
17. Takanaka F et al. (1993) A study of the geo-pollution by chlorinated organic compounds in Kujyukuri plain. In: The proceedings of Geological Society of Japan, centennial academic meeting. Tokyo, pp 34–35
18. Tochigi Prefecture (1994) White paper of environment. Utsunomiya, p 208
19. Japan Water Pollution Research Association (1986) The report to environment agency: research for groundwater protection. Tokyo, pp 107–121
20. (1993) Annual report of Toyama Prefecture Center for Environment, Vol 21. Toyama, p 35
21. (1991) Annual report of Yamaguchi Prefecture Institute for Public Health and Environment, Vol 34. Yamaguchi, p 73
22. Uji-ie A et al. (1993) Groundwater pollution in Miyagi Prefecture, annual report of Miyagi Prefecture Center for Public Health and Environment, Vol 11. Sendai, pp 104–107
23. Yonezawa City (1994) Environment of Yonezawa City. Yonezawa, p 40
24. Gosen City (1994) Environment of Gosen City. Gosen, pp 30–34
25. The National Institute for Environmental Studies (1988) Proceedings of 4th symposium on behavior and characteristics of hazardous compounds in soil and groundwater environment. Tsukuba, pp 73–74
26. Niigata Prefecture Research Laboratory of Public Health and Environment (1988) Report behavior and decomposition of chlorinated organic compounds in environment. Niigata
27. Ito K (1992) Pollution by trichloroethylene in Odawara City. In: Proceedings of 2nd workshop of groundwater pollution and its prevention. Tokyo, pp 116–121
28. (January 1, 1991) Mayor demands compensation for groundwater pollution to company. Nishi-Nihon Shimbun
29. Narashino City (1993) White paper of environment. Narashino, pp 31–42
30. (March 15, 1992) Invitation plan of factory includes possibility of groundwater pollution. Chunichi Shimbun
31. Chiba City (1993) White paper of environment. Chiba pp 183–194
32. (May 14, 1992) Pollution source is Nippon Kogyo. Kanagawa Shimbun
33. Fujieda City (1993) Environment of Fujieda. Fujieda, pp 90–96
34. Fujinomiya City (1992) Pollution in Fujinomiya. Fujinomiya, pp 72–82
35. The National Institute for Environmental Studies, op. cit.
36. Ichihashi, op. cit.
37. Yokohama City (1993) White paper of environment. Yokohama, pp 68–69
38. Nagano City (1993) White paper of environment. Nagano, p 182
39. Fujieda City, op. cit.
40. Gotenba City (1994) Situation of groundwater pollution at Suginazawa area. Gotenba
41. Kobayashi, op. cit.
42. Taniyama M et al. (1990) Groundwater contamination by chlorinated organic compounds in Tsuchiura city, report of Hydrology Experiment Center of Tsukuba University, No 14. Tsukuba, pp 49–57
43. Yamaguchi Prefecture (1993) White paper of environment Yamaguchi, pp 75–77
44. (August 10, 1990) Prefecture ordered Cannon factory. Shimotsuke Shimbun
45. Chuman N (1992) Groundwater contamination with chlorinated solvents in the southern Fukushima Basin, Northern Japan. The annual report of synthetic study at Fukushima University, nature and humanization, No 3. Fukushima, pp 19–30
46. Suzuki Y et al. (1993) The remedial operation for Geo-pollution site at the Kururi–Chiba geo-pollution site. In: Geological Society of Japan (ed.) The proceed-

ings of the third symposium on geo-environment and geo-technics. Tokyo, pp 21–25

47. Narashino City, op. cit.
48. (July 20, 1991) Groundwater pollution at Tsuru City. Yamanashi Nichi-Nichi Shimbun
49. Tonokai K (1993) Reclamation of groundwater and characteristic of groundwater contaminated in Takatsuki City. In: Geological Society of Japan (ed.) Abstracts, the symposium on geo-pollution. Tokyo, pp 97–108
50. (February 8, 1991) President of pharmaceutical company negotiates with citizens, groundwater pollution in Amagi. Nishi-Nihon Shimbun
51. Kashima City (1992) White paper of environment. Kashima, pp 60–82
52. Yoshioka M (1986) Case study of acrylonitrile spill accident. Report of Hyogo Prefecture Research Laboratory of Environment, Vol 18. Kobe, pp 113–117
53. (March 10, 1991) Well of chemical factory at Fujikawa is polluted by carcinogen. Shizuoka Shimbun
54. Matsudo City (1993) Situation and protection of environment. Matsudo, pp 95–107
55. Yoshioka S et al. (1986) Survey of organic compounds of groundwater in Kawasaki City, annual report of Kawasaki City Environment Research Laboratory, Vol 19. Kawasaki, pp 42–47
56. Japan Water Pollution Research Association (1990) The report to environment agency: research for groundwater protection. Tokyo, pp 8–39
57. Shinano-Mainichi Shimbun (1991) With water. Shinano-Mainichi Shimbun, Nagano, pp 31–33, 154–167
58. Ex Metropolitan Institute (1992) Investigation Survey of Appropriate Management Measures for Landfill Sites. The report to environment agency. Tokyo, pp 156–162
59. Hamura City (1992) Report of old landfill site. Hamura, p 163
60. See recent volumes of Man and Environment
61. (November 28, 1962) Landfill site of industrial waste has polluted well? Citizens of Tago City file a suit to stop operation. Chiba Nippo
62. Kurata E (1993) Illegal dumping at Iwaki. Indust 8:8–13
63. Kashiwa City, op. cit.
64. (January 13, 1993) Massive dumping of waste oil. Kobe Shimbun
65. (September 6, 1993) Illegal dumping at Saga. Nishi-Nihon Shimbun
66. Hearings on National Defense Authorization Act for fiscal years 1992 and 1993. Statement by Richard Ray, Representative from Georgia, Chairman, Environmental Restoration Panel. In: HR 2100, before the Committee on Armed Services House of Representatives, 1991. [H201–7, 27], p 1021
67. IARC (1995) Summary and evaluation 63. pp 75, 159
68. Lagokos S et al. (1986) An analysis of contaminated well water and health effects in Woburn, Massachusetts. J Am Stat Assoc 81:583–596
69. Society for Groundwater Protection (1993) Introduction to groundwater. Hokuto Shoten, Tokyo, pp 118–121
70. Shiga Prefecture (1994) White paper of environment. Ohtsu, p 98
71. Fukui Prefecture (1992) Findings of survey on quality of public water area and groundwater. Fukui, pp 88–89
72. Narashino City, op. cit., and Chiba City, op. cit.
73. Tochigi Prefecture, op. cit.
74. Kanagawa Prefecture (1993) Findings of survey on quality of groundwater. Yokohama, p 15
75. Nirei H et al. (1994) Geo pollution unit. J Geologic Soc Jpn 100:425–435
76. Administrative Inspection Bureau of the General Affairs Agency (1994) For protection of public water area and groundwater. Printing Bureau, Ministry of Finance, Tokyo, p 46
77. Professor Chuman of Fukushima University has suggested these four points

78. Administrative Inspection Bureau of General Affairs Agency (1994) op. cit. pp 43–45
79. Morita H (1992) Situation of groundwater pollution in Japan. In: Proceedings of 2nd workshop on groundwater pollution and its prevention. Tokyo, p 262
80. Sakai Y et al. (1995) Research-measure system of geopollutions. In: Geological Society of Japan (ed.) symposium: the responsibility of geo-pollutions. Tokai University Press, Tokyo, pp 57–76
81. Ohtsuka T (1994) Allocating cleanup costs of soil contamination in metropolitan area. Jurist, No 1040, p 97
82. Ibid.
83. Nirei H (1995) From several problems of geo-pollution sites to clean up. In: Geological Society of Japan (ed.) symposium: the responsibility of geo-pollutions. Tokai University Press, Tokyo, pp 11–13
84. (1993) Indust Vol 8, No 9:22
85. Kitamura Y (1993) Limit of administrative response and legal enforcement. *Jichi-Kenkyu* 69:82
86. Ohtsuka T, op. cit., p 100
87. Ohtsuka T (1994) Allocating cleanup costs of soil contamination. Waste Manage Res 5:18
88. Ohtsuka T (1994) Jurist, p 101

Chapter 7
Itai-Itai Disease and Countermeasures Against Cadmium Pollution by the Kamioka Mine*

1. Introduction

Osteomalacia, softening of the bones as a result of renal tubular dysfunction, is now commonly known as Itai-Itai disease. Osteomalacia occurs in women who have borne several children. During the final stages of osteomalacia, when patients can neither stand, walk, nor talk, all they say is *itai-itai* (Japanese for "It hurts! It hurts!"), hence its common name. In one of the worst cases of the disease, a patient lost 30 cm in height as a result of pressure fractures of the vertebrae; and another patient suffered many fractures, 28 in the ribs alone.

The issue was recognized as serious when cases of Itai-Itai disease and renal tubular dysfunction were found among residents in a number of cadmium-polluted areas throughout Japan, most seriously in the district of Toyama near the Jinzu River, but also in Tsushima, Nagasaki Prefecture, the Kakehashi River basin in Ishikawa Prefecture, and the Ichi River basin in Hyogo Prefecture.

Japanese people have higher levels of daily cadmium intake and a higher concentration of cadmium in the renal cortex than any other group of people in the world. Because high cadmium levels have been found in rice grains, it is reasonable to suppose that this is the route by which cadmium enters the human body, from which it follows that the problem of toxic pollution by cadmium will not be confined to the locality in which the cadmium was originally dispersed into the soil.

In 1968, victims of Itai-Itai disease living in the Toyama district brought a case against the main polluters (the Mitsui Mining and Smelting Company). After the court had reached a decision and made a judgment against Mitsui, the victims and the corporation held negotiations and agreed on three points: (1) that compensation should be paid to the sufferers of Itai-Itai disease; (2) that compensation should be paid to those whose fields and crops had been polluted, as well as compensating them for the detoxification of contaminated farm soil; and (3) that pollution-prevention systems should be installed at the Kamioka mine and refinery. This last requirement has proved to be the most effective of the measures taken, because the company agreed

* Originally published in *Environmental Economics and Policy Studies* (Springer, Tokyo, 1999) 2:215–229. Originally co-authored with A. Hata and H. Tonegawa.

to allow the victims, their lawyers, and interested scientists to survey and inspect the facilities without hindrance and at the company's expense.

Subsequently, sufferers, lawyers, and scientists have carried out an annual inspection of the Kamioka mine and refinery every year since 1972. In particular, the unhampered on-site inspections, the collection of data and disclosure of information about the type of material involved in the process, updated reports on the quality of process water and the conditions of drainage and emission, and the state of the soil, groundwater, and by-products have all been powerful measures in preventing further pollution.

This chapter, which is based on the reports of those inspections, considers the following issues: (1) the history of the disturbances to the environment caused by the Kamioka mine; (2) the Itai-Itai disease suit; (3) the pollution-prevention measures and methods used by the Kamioka mine and refinery; and (4) the reduction of cadmium pollution in the Jinzu River.

2. History of Environmental Damage Caused by the Kamioka Mine

In general, a mining operation consists of three stages: (1) the mining process; (2) the ore-dressing process; and (3) the refinery process. All three stages have an effect on the environment. Although the metal-mining industry has played an important role in the modernization of Japan, the extensive development of the industry has also been responsible for many instances of environmental disturbance and damage.

In 1874, the Mitsui Corporation bought the Kamioka mine, in 1886 it introduced modern technology, and in 1889 it brought all the Kamioka mine operations under the umbrella of one administration. Within 1 year (in 1890), the local air was sufficiently polluted by White's rotary furnaces for the residents of Kamioka Town to lodge a complaint: this was the first instance of a formal protest by residents.

Since 1905, the main work of the pit has been the mining of zinc ore, which, from the beginning, required the development of technology to treat the poorer-quality ore and to refine its dressing process. In 1909, Mitsui Corporation installed Potter's preferential flotation system to cope with its ore-dressing process. As means to treat the poor zinc ore were developed, the amount of ore increased, and the preferential system resulted in finer tailing particles. The tailing slurry was thrown into the Jinzu River and was then precipitated to the rice paddies of the Toyama District. At the same time, the first cases of people suffering from Itai-Itai disease were reported.

After 1931, the organization representing the farmers, peasants, and fishermen of the Jinzu River basin repeatedly asked Mitsui Mining Corporation to furnish proper facilities to prevent pollution; in response, Mitsui built its first tailing dam in the Shikamadani Valley (1931). The period during the war (1930–1945) saw a fourfold increase in the production not only of zinc, but also of zinc waste, which meant that more extensive areas of land in the Toyama District in and around the Jinzu River basin became contaminated.

Although the main aims of the "innovations" introduced after World War II (mass production and the saving of manpower) were designed to effect speedy restoration of the Kamioka mine and stimulate high economic growth, it took nearly 10 years for the Kamioka mine to reach prewar production levels.

As time went on, further innovations were introduced. In 1966, a significant change was made in the method of dressing ore: instead of the direct bulk differential flotation process it had used until then, the company adopted integrated bulk differential flotation. This process markedly increased the amount of ore-dressing treatment undertaken, which, in turn, led to a massive increase in tailing waste, with particles far finer than before. Because insufficient measures had been taken to prevent the random disposal of tailing waste (an earlier example of which was the collapse of the tailing dam at Wasabo in 1956), this inevitably led to a severe increase in pollution emanating from the Kamioka mine.

Although installation in 1944 of a German zinc electrolysis plant was intended to assist the recovery of cadmium, the period of high economic growth during the 1960s saw the production of vastly increased amounts of treated ore. The total loss of cadmium from the main processes of the Kamioka mine and refinery reached its highest recorded levels in 1970.[1]

3. Itai-Itai Disease Suit

When, after the end of World War II, Dr. Noboru Hagino, a physician in Fuchu Town, Toyama Prefecture, reopened his private clinic in the town, he was alarmed by the number of patients who came to see him suffering from particularly severe and painful attacks of osteomalacia. It was not until 1955, however, that he and Dr. Minoru Kohno alerted the general public to the nature of Itai-Itai disease.

Although over the next 2 years, a number of reports were published insisting that the osteomalacia was simply the result of vitamin deficiency, Dr. Hagino continued to investigate the possibility that other causes were responsible. He strongly suspected that the hazardous outflow from the Kamioka mine was a likely source of the disease because the area of agricultural land contaminated by the Kamioka mine perfectly overlapped the area from which the afflicted patients came.

In 1961, after Dr. Jun Kobayashi had published his data on the cadmium content of rice obtained from the contaminated land, Dr. Hagino and Dr. Kin-ichi Yoshioka published a report that set out to show that cadmium played an etiological role in the development of Itai-Itai disease. Two years later, in 1963, the Ministry of Health and Welfare instituted a study group to inquire into the disease. The group published its report in 1967, concluding that, in conjunction with other factors such as nutritional deficiencies, cadmium was one of the causes of Itai-Itai disease.[2]

The victims were not happy with this report, and that same year they set up an organization (Task Force on Itai-Itai Disease) to ensure that the real culprit, cadmium, should be publicly acknowledged. In 1967, the task force entered into negotiations with the Mitsui Mining and Smelting Company, but nothing fruitful emerged from the discussions.

It appeared to the task force that the political and administrative approaches were likely to remain closed, so the task force decided, as a last resort, to take legal action. Supported by local residents and local government (town), more than 300 lawyers from all over Japan were organized to present the suit, which they filed against the

Mitsui Corporation, basing it on Article 109 of the Mining Law, which covers a polluter's strict liability.[3,4] Soon after the appeal was brought to court in 1968, the Ministry of Health and Welfare declared "Itai-Itai disease is caused by chronic cadmium poisoning during times of pregnancy, lactation, or when there is an imbalance of internal secretion, aging, a lack of calcium, etc."

In 1971, the Toyama district court handed down a decision that accepted the victims' claims as proven. Although Mitsui Mining and Smelting Company appealed the decision to the Nagoya High Court and disputed the causality, the verdict for the victims was upheld. On the basis of this decision, Mitsui Corporation made reparations to more than 200 individuals suffering from the disease.[5–7]

4. Pollution Prevention After the Itai-Itai Disease Suit

Since 1972, the sufferers, lawyers, and interested scientific parties have carried out inspections of the Kamioka mine and refinery. The first two inspections revealed that: (1) the total amount of cadmium being discharged into the Jinzu-Takahara River was 35 kg/month (1972); (2) the total amount of cadmium being discharged into the air was 5 kg/month (1972); (3) although both Mitsui and the agents for the victims checked the same samples, the data did not always coincide, so it could not be accepted as an accurate measurement; and (4) discharge from abandoned mines had extensively polluted the Jinzu-Takahara River.

Five working groups were organized to carry out remedial measures:

1. To control the effluent from the Kamioka mine and refinery (Kyoto University);
2. To control smoke emissions from the Kamioka refinery (Nagoya University);
3. To monitor the balance of materials at the Kamioka refinery (Tokyo University);
4. To monitor the sedimentation and outflow of heavy metal into the Jinzu-Takahara River system (Toyama University); and
5. To monitor the structural stability of the tailing dams at the Kamioka mine (Kanazawa University).

The Kamioka mine, the largest zinc mine in Japan, is located midway along the northwest coast of Japan's crescent-shaped main island, Honshu (Fig. 7.1). Two plants use a bulk-differential flotation process to refine the crude ore from the mine to lead, zinc, copper, and graphite concentrates. The lead concentrates are roasted and sintered, after which the coarse sinter is fed into a blast furnace. The crude lead is refined in electrolysis cells. The by-products of the lead refinery are gold, silver, and bismuth. The zinc concentrates are roasted into zinc calcine, which is then leached and refined in electrolysis cells. The by-products of the zinc refinery are sulfuric acid and cadmium.

As a result of closure of its Mozumi mine, the Kamioka Mine and Smelting Company has recently changed the material used in its lead refinery processes from lead to lead batteries, and it has begun to import zinc ore from the Huanzala mine in Peru.

The effluent from the Kamioka mine is managed as follows. The average amount of water drawn from the mine is approximately 100000 m³/day. The mine water is divided into two grades: clear water and muddy water. The tailing slurry from the ore

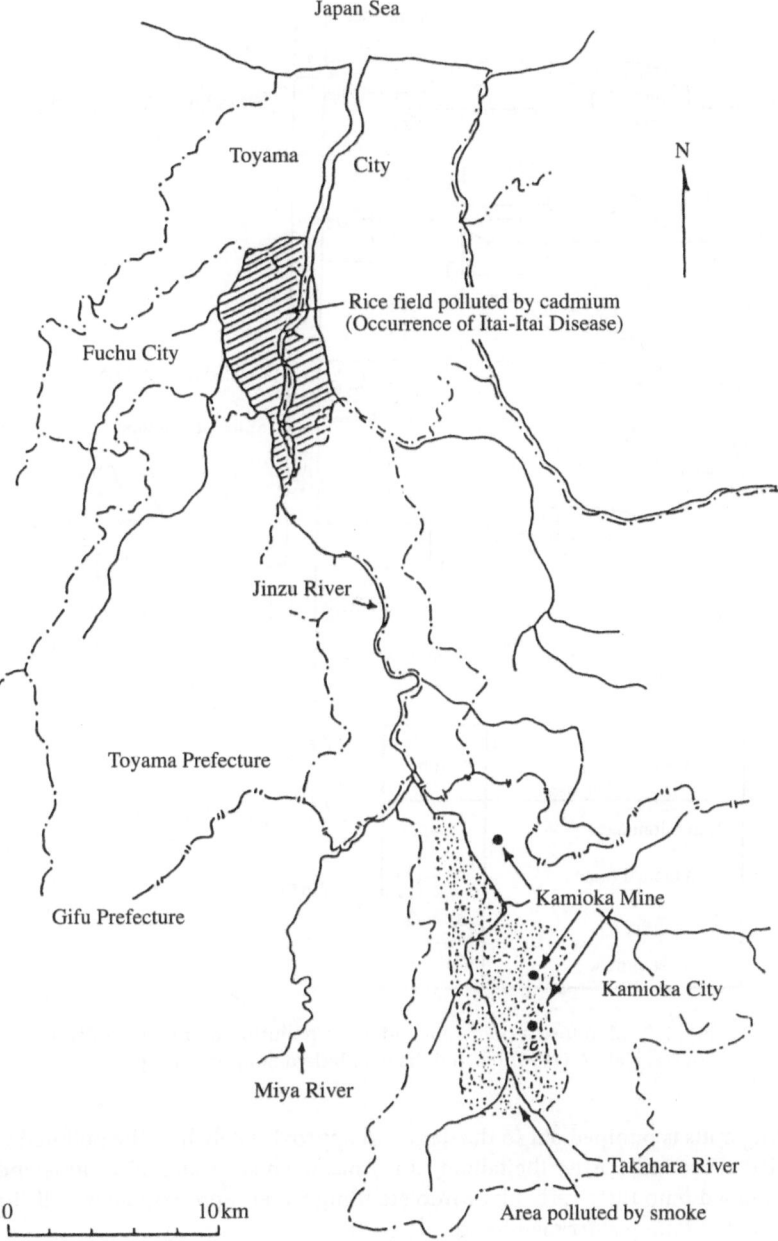

FIG. 7.1. Area polluted by the Kamioka mine and refinery
Source: Kurachi M, Tonegawa H, Hata A (eds) (1979) Mitsui Company and Itai-Itai disease.
Ohtsuki Shoten, Tokyo

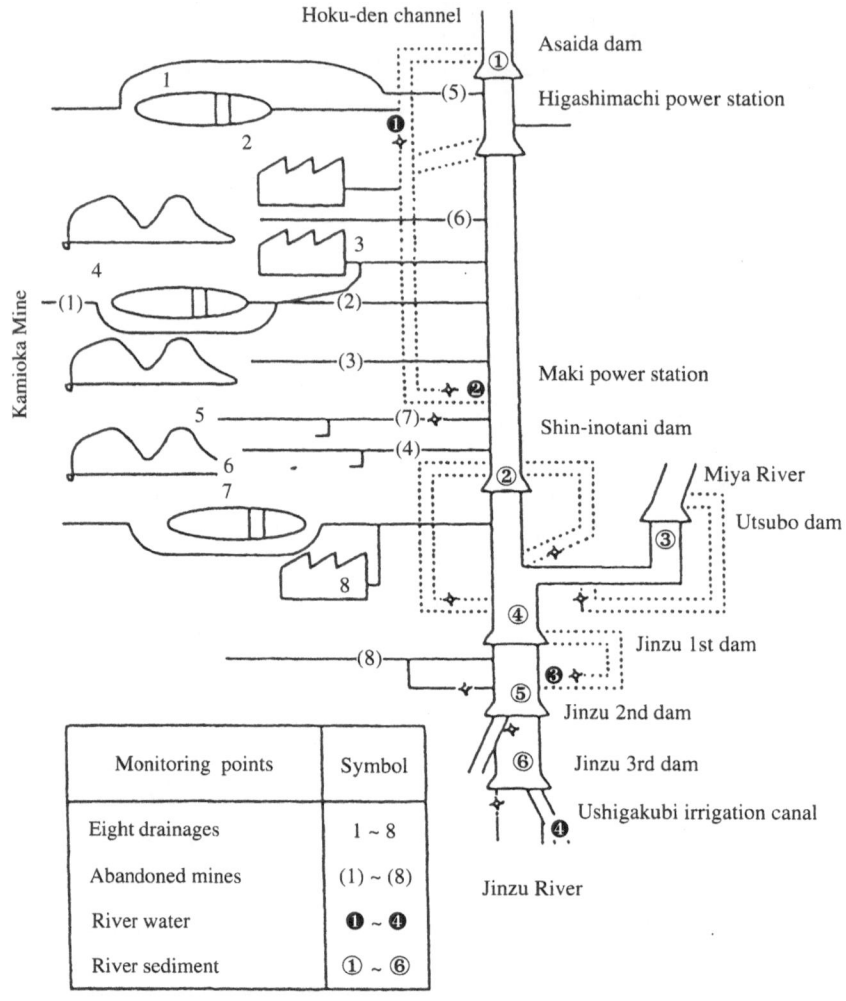

FIG. 7.2. Monitoring points for cadmium pollution on the Jinzu River
Source: Hata A (1996) Itai-Itai disease. Jikkyo Shuppan, Tokyo

dressing mills is pumped out to the dam and approximately half the polluted water is recycled at the mills. After the tailing slurry has been separated into slime and sand, it is dumped onto the riverbank or into standing water; approximately half the effluent from the tailing slurry is reused.

Drainage of effluent and waste water from the Kamioka mine and refinery is monitored at eight authorized checkpoints: (1) the Wasabo tailing dam; (2) the Rokuro zinc plant; (3) the Shikama ore dressing, lead, and sulfuric acid plant; (4) the Shikamadani tailing dam; (5) the Atotsu audit mouth; (6) the Otsuyama audit mouth; (7) the Masutani tailing dam; and (8) the Mozumi ore-dressing plant (Fig. 7.2).

The total amount of effluent water is 110000 m³/day. Thanks to the separation of clear water from muddy water, the recycling of process water at the ore-dressing plant

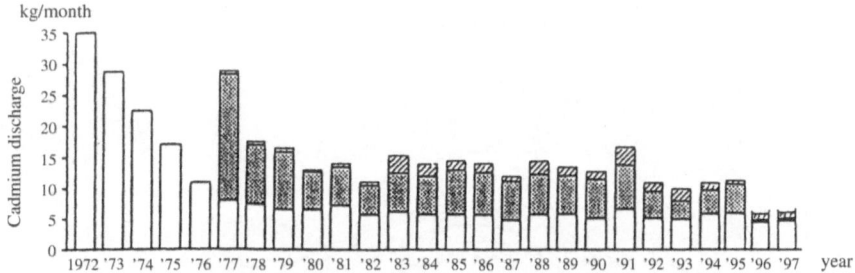

FIG. 7.3. Transition of cadmium discharge by the Kamioka mine and refinery. *Cross-hatched bars*, discharge from abandoned mines; *shaded bars*, discharge from the Hoku-den channel; *open bars*, discharge from the eight drainage outlets

and the electrolysis cells, and an improvement in the process of wastewater treatment, the total amount of cadmium discharged has been reduced from 35 kg/month (1972) to 5 kg/month (1997; Fig. 7.3). At the same time, recycling the process water has brought about a saving in both materials and reagents, whereas improved dust collection has reduced the total amount of cadmium discharged in smoke from more than 5 kg/month in 1972 to 0.4 kg/month in 1997. Apart from any unpredictable consequences that might result from a strong earthquake, scientists have ensured the stability and structural safety of the tailing dam.

5. Reduction of Cadmium Pollution in the Jinzu River

The Jinzu River, called the Takahara River at its source, arises in the Hida Mountains, an area affected by heavy winter snow. The river flows northward toward the Japan Sea (Fig. 7.1). The background cadmium level of the upper reaches of the Jinzu–Takahara River, unpolluted by the Kamioka mine, is never more than 0.1 ppb. In contrast, the cadmium level downstream of the Kamioka mine can be as high as 10 ppb.

The total amount of known cadmium discharge from the eight authorized drainage points is approximately 5 mg/s and 1–3 mg/s from the abandoned mines. At the Shin-Inotani Dam, however, which is located downstream of the Kamioka mine, the cadmium discharge into the river water can be as high as several tens of mg/s.

Attention was therefore paid to locating other likely, although at that time unconfirmed, sources of pollution. The chief suspect was an underground tunnel that runs underneath the zinc electrolysis plant from a nearby power station. Measurements taken in 1977 revealed that the cadmium discharged from this channel accounted for as much as 70% of the pollutants from all outlets. At the same time, the quality of the groundwater from the drilling wells at the Rokuro electrolysis plant suggested that the plant might be an even more serious source of contaminants. Thorough investigation into the causes of pollution linked to the underground tunnel and measures taken to counteract groundwater pollution have been both the most important and the most difficult items on the agenda for 20 years.

When the underground channel was inspected, it was found that groundwater with a high cadmium content issued from: (1) an open sewer at Rokuro; (2) the drainage

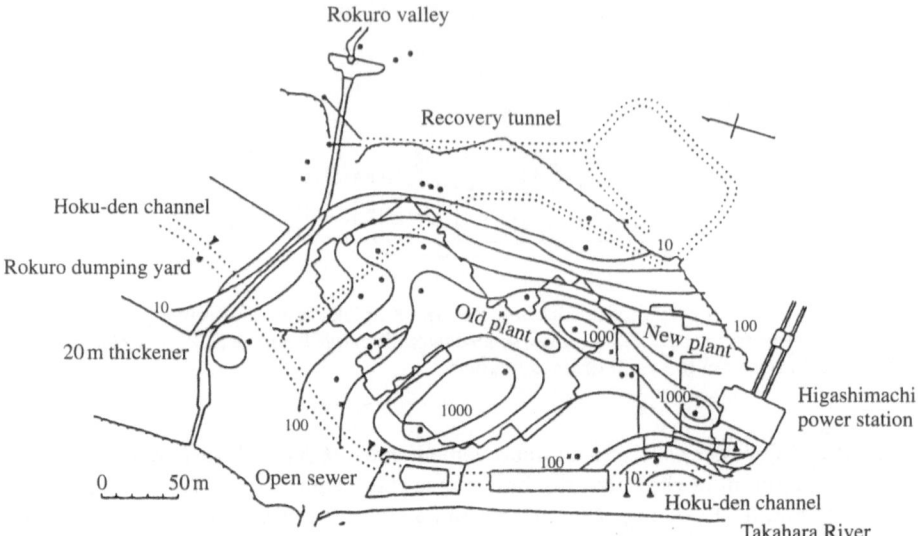

FIG. 7.4. Distribution of cadmium concentration (ppb) in the groundwater at the Rokuro zinc electrolysis plant, March 1994. *Triangles*, spring water; *circles*, test drilling; *X*, recovery well Source: Hata A (1996) Itai-Itai disease. Jikkyo Shuppan, Tokyo

pond of the Higashimachi power station; and (3) the Rokuro dumping yard (Fig. 7.4). The Kamioka Mining and Smelting Company therefore agreed to take action to protect the groundwater system from pollution: (1) to prevent leakage of contaminated groundwater into the channel; (2) to pump up the contaminated groundwater through a recovery well; (3) to cut off the infiltration of Rokuro Valley water into the ground; (4) to analyze the quality of the groundwater at the site of the Rokuro electrolysis plant; and (5) to repair or renew the floor and tanks of the Rokuro plant (Fig. 7.4).

Although as a result of these measures cadmium discharge from the channel fell from 21 kg/month in 1977 to 7 kg/month by 1980, during the period 1980–1990 the cadmium discharge did not change (Fig. 7.3). Hence, in 1990, the Itai-Itai disease sufferers requested that the old electrolysis plant be rebuilt. The following year, 1991, the company spent ¥6 million to reconstruct the electrolysis plant with the aid of Belgian technology.

During the reconstruction, it was discovered that the soil under the new plant, which was built alongside the old plant, was also severely contaminated with cadmium and zinc. The company therefore drilled cores at 15 sites to survey the contaminated soil and groundwater. On the basis of this survey, the company has extended the recovery tunnel, which, for the purpose of pumping up the groundwater, was excavated along the eastern margin of the polluted area.

Use of the tunnel and the operation of the new plant have brought about a reduction of approximately 2–3 kg/month in cadmium discharge. After 1995, the company spent ¥90 million to pump up the contaminated groundwater in the Hokuriku Electric Power Company channel. As a result of these measures, the cadmium discharge

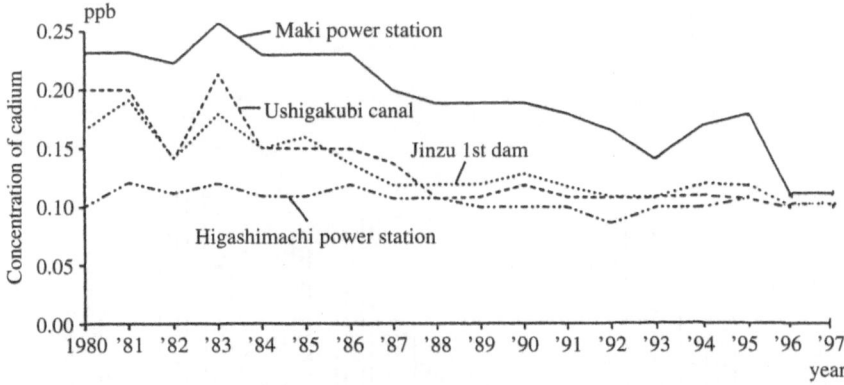

FIG. 7.5. Transition of cadmium concentration in Jinzu River water

from the Hokuriku Electric Power Company channel dropped 0.3 kg/month after 1996 (Fig. 7.3). The mean concentration of cadmium in the river water at the Jinzu third dam fell from 1.5 ppb in 1968 to 0.16 ppb. in 1977. After 1980, the mean concentrations of cadmium in river water at the Jinzu first dam and the Ushigakubi irrigation canal have been 0.1–0.2 ppb, which is approaching the background level of cadmium found in the non-polluted area since 1988.

Because the cadmium discharge from the Hokuriku Electric Power Company channel fell to approximately 0 kg/month after 1996, the mean concentration of cadmium in river water anywhere along the Jinzu River was at 0.1 ppb (Fig. 7.5). The mean concentration of cadmium found in sediment at the Jinzu first dam fell from a level of 16–18 ppm in 1968 to 1.39 ppm in 1976. Although since 1986 the level has been less than 1 ppm, this figure is slightly higher than that found in the non-polluted area (less than 0.5 ppm).

6. Conclusions

If we consider that the restoration of land to its original condition entails, as a corollary, that the same pollution damage should never occur again, source prevention measures and soil pollution-prevention measures are two of the main pillars supporting the recovery of land damaged by pollution. In this regard, Kamioka Mining and Smelting Company source prevention measures, based on the pollution prevention agreement, have achieved a remarkable success by obtaining a natural background level of 0.1 ppb cadmium outflow to the downstream area. These efforts should be highly valued.

Although we have sought to make it clear that Mitsui Mining and Smelting Company has, to a certain extent, managed to evade some of its obligations, the court decision and the agreements made between the victims and the company have nevertheless made it possible to control, and subsequently reduce, the physical and eco-

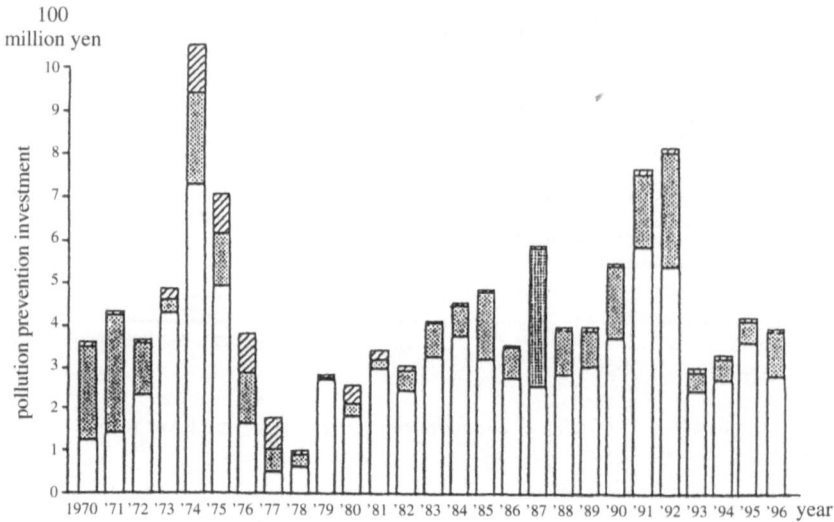

FIG. 7.6. Transition of pollution prevention investment in the Kamioka mine and refinery. *Cross-hatched bars*, treatment of abandoned mines; *shaded bars*, treatment of effluent smoke, *open bars*, treatment of effluent water

nomic damage caused by cadmium pollution emanating from the Mitsui Kamioka mine. We draw attention to four contributory factors: (1) the geographical and administrative separation of the mine area and the homes and farms of the sufferers, which has meant that "company town" problems have been avoided (as when powerful companies "lean on" their employees to conform or risk losing their jobs and their homes); (2) a firm tie-up between the victims, who are typically rice farmers, and the local (town) government to assist the sufferers; (3) a more serious undertaking over the past 25 years by Mitsui Mining and Smelting Company (Fig. 7.6) to listen to the victims and their lawyers and to pay attention to their requests; and (4) an improvement in Mitsui Kamioka Mining and Smelting Company's attitude to the environment and an increased willingness to look after it (because the company's size gives it the technical means to do what it has a moral obligation to undertake), and to carry out the necessary "downsizing and restructuring" of its own plant. Therefore, the outlook is not entirely black.

References

1. Kurachi M, Tonegawa H, Hata A (eds) (1979) *Mitsui Shihon to Itai-Itai Byo* [Mitsui Company and Itai-Itai disease]. Ohtsuki Shoten, Tokyo
2. Research Group of Itai-Itai Disease (1967) Research Report on so-called Itai-Itai Disease, Kanazawa
3. Gresser J, Morishima A (1981) Environmental law in Japan. MIT Press, Cambridge MA, Chapter 3

4. Nogawa K et al. (eds) (1998) Advances in the prevention of environmental cadmium pollution and coutermeasures: proceedings of the international conference on Itai-Itai disease, environmental cadmium pollution and countermeasures, 1998, Toyama. Eiko Laboratory Kanazawa
5. Lawyers for Itai-Itai Disease Sufferers (1971–1974) Documents of Itai-Itai disease suit, Vol 1–6. Sogo Tosho, Tokyo
6. Coalition of Sufferers by Cadmium along the Jinzu River Basin and Investigation Committees (1978) Pollution prevention at the Kamioka mine, 1978, Toyama
7. Kamioka Mining and Smelting Company (1979–1997) Annual report of pollution prevention, 1979–1997. Kamioka

Chapter 8
Accumulated Pollution and the
PPP—Mainly Heavy Metal Pollution*

1. Introduction

When the OECD first formulated its policy that an industrial polluter should pay for any pollution caused by heavy metals and PCB in the soil or water, the policy was mainly targeted at the prevention of pollution in the first place, at which stage the cost of any preventative measures would be absorbed as a preliminary cost by the industrialist: "the internalization of the external cost" (the Polluter-Pays Principle: PPP).

When Japan, taking its lead from this policy, initiated its own unique form of the PPP, it extended the original concept to also take into account the payment of retrospective costs for pollution cleanup, and compensation for any long-term accumulation of heavy metals and PCB; this has been defined as "the retrospective environmental cost". Thus, Japan's PPP has the character of a legal liability, and specifies that the polluter has to pay for the cost of cleanup and compensation.[1]

However, since the establishment of the US Superfund system (CERCLA/SARA) in the 1980s, the application of the PPP to cases of accumulated pollution is no longer exceptional. Therefore, we have to reinvestigate the theory for the actual conditions of accumulated pollution when the PPP can be applied. In this chapter, the author aims to examine these issues afresh, and will make a precedent of cases of accumulated pollution caused by heavy metals and the way in which, in Japan, the PPP has been applied.

2. Law Concerning Companies Bearing of the Cost of the Public Pollution Control Works and the Agricultural Land Soil Pollution Prevention Law

Of 14 pollution-related bills approved in December 1970, during "the Environmental Diet", the 64th Diet of the Lower House, the Agricultural Land Soil Pollution Prevention Law was enacted against the background of Itai-Itai disease, which was the result

---

* Originally published in *Economic Journal of Hokkaido University* (Sapporo, 1998) 27:11–34. With the permission of the University.

of cadmium poisoning, and the nationwide soil pollution problem that has afflicted agricultural land. In 1968, the case of Itai-Itai disease was brought before the court, and the Ministry of Health and Welfare was forced to acknowledge that Itai-Itai disease had, indeed, been caused by cadmium. Sufferers have, at last, been certified as victims of Itai-Itai disease. The revision of the law added soil pollution to six kinds of pollution that had, by then, been recognized. In order to complement the addition, the Agricultural Land Soil Pollution Prevention Law was enacted. The reason for its enactment was that, against a background of rice contaminated by cadmium and because of a concern for public health, the prevention of land soil pollution and the elimination of existing pollutants had become recognized as unavoidable issues.[2]

During Diet discussions of the Committee of Agriculture, Forestry and Fisheries, members emphasized the close relationship between "a human being's health" and "rice for food" because all Japanese eat rice. Nevertheless, the problem of heavy metals other than cadmium and the issue of soil pollution of non-agricultural land remained unaddressed. Heavy metals, such as copper, zinc, and arsenic, were, one by one, added to the list of recognized contaminants. Although the direct relationship between pollutants and human health was stressed, no mention was made of the conservation of the whole ecosystem. Soil pollution of non-agricultural land was referred to only in a supplementary resolution because ". . . human health cannot be damaged unless a human being eats a sufficient quantity of contaminated agricultural products".[3]

Thus, the discussion focused on the relationship between human health and agricultural products. The next crucial point, of who should bear the expense of any remedial action, was settled during discussion of the Law Concerning Companies Bearing of the Cost of the Public Pollution Control Works in the same 64th Diet, with an emphasis, as a whole, on the need ". . . not to place a burden on the farmers"; after the polluter's payment had been deducted, the remaining expenses were to be paid for by public money.

As to the allotment of the PPP, both the Ministry of International Trade and Industry (MITI) and the Ministry of Health and Welfare spent as long as 3 years attempting to determine a suitable ratio, but in vain.[4] It was only when the successor to the Environmental Pollution Control Headquarters, the Environment Agency, was established that the differences of opinion were finally recognized and balanced. According to the Ministry of Health and Welfare's interim report[5] regarding the work of treating accumulated pollutants, the agricultural land soil cleanup business set a minimum rate of three-quarters of the whole expenses as the share to be borne by the private sector. However, Article 4, Item 1 of the newly established Law Concerning Companies Bearing of the Cost of the Public Pollution Control Works states ". . . the total amount borne by the company should correspond to [the degree of recognition of] how much the company's activities have affected the cause of the pollution". Item 2 of the same law gives three reasons for reducing the amount of costs levied on the companies: (1) "the function other than pollution control"; (2) "the degree of pollution needing pollution control"; and (3) "circumstances covering the period of contamination by environmental pollution as these relate to the relevant work of pollution control".

Article 7 of the same law states that ". . . in cases where the cost of reducing pollution cannot be calculated under the regulation", it should be possible to establish a standard amount multiplied by the amount of money mentioned in Article 4, Item 1, by an approximate ratio of half to three-quarters. Due to hastily enacted regulations designed to reduce the various kinds of cost to be borne by the polluter, the resultant rate of reduction (of costs) was fixed at a maximum of three-quarters. Once paid, the polluter was allowed to calculate his costs as necessary temporary expenses or as a financial loss. Although the "reduction of costs" item incurred much criticism, even in the Diet, the law was finally passed without any amendment so as not to further burden the farmer.[6]

In the draft bill, there is no mention of "the rational use of agricultural land", which nevertheless appears as Article 1 (the purpose) of the Agricultural Land Soil Pollution Prevention Law. This was a newly added item, and there was additional concern that it would stimulate the conversion of contaminated land to non-agricultural uses.[7]

The Environmental Pollution Control Headquarters commented on Article 4, Item 1 of the Law Concerning Companies Bearing of the Cost of the Public Pollution Control Works in these terms:

"It is not so difficult to calculate the rate of the cause of pollution as it relates to the relevant work of pollution control because almost nothing is to be considered in relation to any other causes of pollution, such as the creation of a green belt or soil cleaning-up of farm land."[8]

However, at first, the Headquarters reacted negatively to what came to be called the "natural pollution" theory—that, for geological reasons, the level of cadmium is higher in certain areas than in other areas. But, in a book published later, *Comment on the Agricultural Land Soil Pollution Prevention Law*, co-edited by the Water Quality Control Bureau of the Environment Agency and the Agricultural Administration Bureau of the Ministry of Agriculture and Forestry, we find the statement that ". . . the share to be borne should correspond to the degree of recognition that the entrepreneur's total company activity has caused the relevant pollution, namely to the exclusion of other indirect causes of pollution and so on". This comment made way for the development of the natural pollution theory.[9]

A request for the reduction of payment because of natural pollution had already been presented by the mining industries at the time, and was mentioned in a report presented by the Mining Council in July 1974, entitled "On How To Carry Out Measures to Prevent Accumulated Pollution in Metal Mining".[10] As a whole, the Agricultural Land Soil Pollution Prevention Law and the Law Concerning Companies Bearing of the Cost of the Public Pollution Control Works, simultaneously enacted in the 64th Diet, were epoch making and, at the time, were the most advanced provisions in the world for ensuring a practical way of covering the expenses of the restoration of agricultural land soil in terms of the PPP.

The soil restoration stipulated by the two laws, however, applied only to agricultural land and, for two reasons, namely human health and security of food, the stipulation did not extend to land soil pollution in general because it lacked any recognition of the need to secure a healthy ecosystem and environment. Consequently, the PPP was hardly adhered to. The principle of "do not burden the farmer" is a common understanding of the two laws and what we can say from the recognition of this background is that the two laws are a prescription to pay for the cleanup expenses out of public funds that have been provided by both the central government and local authorities, which results in payment for the cleanup of accumulated pollution with tax-derived subsidies.

FIG. 8.1. Areas of the Agricultural Soil Pollution Control Project in Japan. Notes: (1) The under-lining of ◖, ▲ shows the pollution caused by the plural specified toxic substancesf; (2) ◔, △ shows the designated area polluted by cadmium and copper, a part of which is canceled the designation

3. Measures Against Soil Pollution as a Result of the Two Laws

The Agricultural Land Soil Pollution Prevention Law led to the designation of soil pollution prevention of land amounting to 6260 ha in 66 districts (Fig. 8.1). There was a total of 39 cases of soil dressing with additional top-soil replacement, in line with the Law Concerning Companies Bearing of the Cost of the Public Pollution Control Works, for the 25 years from May 1971, when the law was first applied, to the end of 1995. The total business expenses amounted to ¥79 billion and the share borne by the entrepreneur (the polluter) was 42.6%. Compared with the number of dredging operations (33) and the entrepreneur's share of the costs (66.6%), based on the same law over the same period, the figures are characteristic of the smaller share borne by the entrepreneur.[11]

However, the company share of the costs for measures for the prevention of soil pollution had originally been relatively high, mostly 75% (see cases 1–6, Table 8.1, from 1972 to 1975). Once soil pollution measures to be undertaken by Takanosu District and the Shinjo and Tokomai Districts of Akita Prefecture had been announced on December 9, 1975, as well as for areas around Ikuno Mining in Hyogo Prefecture, announced in 1976, the entrepreneur's share of the burden began to drop: for example, that for Akita Prefecture went down to 44.3%, whereas in Hyogo Prefecture the entrepreneur's share of the costs dropped dramatically from 75% (during the first application of the measures) to 59.5%. Because, in the case of Akita Prefecture, there was no direct polluter, the ratio of the contribution to the degree of pollution was calculated for the first time (see Article 4, Item 1) and there was a still further reduction of the approximate ratio by half because of its being "a unique case of geological deposit".[12] Two cases in Akita Prefecture were given prominence by the Japanese Mining Association, which claimed ". . . our insistence [that we were not the guilty party] was finally accepted, at least in part".[13]

In the case of Ikuno Mining in particular, what has been called the natural pollution theory was given publicity because it was impossible to deny completely that the natural pollution had been caused by cadmium.[14]

In the Agricultural Land Pollution Prevention Measures for the Jinzu River Site (first application of the measures), announced in February 1980, the company's share of the costs was set at 33.15%, while the rest of the cost, 66.85%, was covered by national and local government funding. With regard to the Jinzu River Site, contaminated with cadmium from Mitsui Metal Kamioka Mining, an area of 1500 ha was designated as a "pollution-prevention measures district": this is equivalent to approximately one-quarter of the total area covered by the Japanese pollution prevention measure, 6080 ha at that time.

Two commitments were undertaken and an agreement was concluded in August 1972 in negotiations between the Tokyo Head Office of Mitsui Mining and Smelting Company and the resident victims, who were well known to have won a full-scale suit in the second hearing of the Itai-Itai disease case. In the commitment on "the soil pollution problem", Mitsui Mining stated: "once the agricultural land soil restoration measures are put into operation, Mitsui Mining will bear, (A) as a polluter, the total amount of expenses for the measures, (B) the expenses for land re-adjustment and so

TABLE 8.1. Application examples of the Law Concerning Entrepreneurs Bearing of the Cost of the Public Pollution Control Works related to Agricultural Land Soil Pollution Prevention

Name	Date	Total Cost (1000 yen)	Entrepreneur's Bearing (1000 yen)	Rate	Ratio of Contribution	Approximate Ratio		Works Period (Years)	Enforcement	Source	Polluter
1 Usui River	72.9.13 (plan changed)	630 000	472 500	75%	100%		Replace soil	1972–74			
	74.2.15	733 300	549 975	75%	100%			1972–74	Gunma Prefecture	Cd	Toho Zinc
	75.2.14 (plan changed)	912 600	684 450	75%	100%			1972–75			
	76.3.19	870 500	652 875	75%	100%	75%		1972–76			
	78.6.15	1 555 716	1 166 787	75%	100%	75%		1972–82			
2 Ikuno Mining	73.7.31 (plan changed)	354 000	265 500	75%	100%		Replace soil	1973–74	Hyogo Prefecture	Cd	Mitsubishi Metal Mining
	74.11.26	428 778	321 584	75%	100%	75%		1973–74			
3 Nakano	74.1.10	131 700	79 020	60%	80%	75%	Replace	1973–74	Nagano Prefecture	Cd	Koshina Optics
4 Kariya	74.1.28	1 500 000	1 123 875	74.9%	99.9%	75%	Replace	1973–74	Aichi Prefecture	Cd	Nippon Denso
5 Nisso Metal Aizu	74.7.3 (plan changed)	273 000	204 750	75%	100%	75%	Replace	1974–75	Fukushima Prefecture	Cd	
	75.3.7 (plan changed)	362 000	271 500	75%	100%	75%		1974–75		Cd	Nisso Metal
6 Tsubo River	76.3.30	406 316	304 737	75%	100%	75%		1974–76			
	75.9.15	56 800	42 600	75%	100%	75%	Replace	1975	Tenmabayashi (Aomori)	Cu	Japan Energy
7 Takanosu Shinjo Tokomai	75.12.9 (plan changed)	344 000	152 478	44.3%	88.65%	50%	Replace	1975–78	Akita Prefecture	Cd	Dowa Mining
	78.2.28	280 243	124 219	44.3%	88.65%	50%					
8 Ikuno	75.12.9	989 290	318 087	32.9%	43.8%	75%	Replace	1975–78	Akita Prefecture	Cd	Dowa Mining
	76.12.17	353 000	210 035	59.5%	79.34%	75%	Replace	1976–77	Ohkochi (Hyogo)	Cd	Mitsubishi Mining

No.	Name	Date			%	%	%	%		Period	Location	Pollutant	Company
9	Kakehashi River	77.8.2 (plan changed)	4 530 000	1 872 249	41.3%	60.76%	68.02%			1977–82			
		79.7.13	5 840 000	2 413 672	41.3%	60.76%	68.02%		Replace	1977–82	Ishikawa Prefecture	Cd	Japan Energy
10	Sasagadani downstream	84.1.27	9 128 720	3 772 899	41.3%	60.76%	68.02%			1984–88			
		78.1.13 (plan changed)	1 117 000	90 477	8.1%	10.8%		75%	Replace	1977–82	Shimane Prefecture	Cd, As	Japan Energy
11	Yoshino River	87.12.11	1 195 365	96 824	8.1%	10.8%		75%		1977–89			
		78.2.15 (plan changed)	3 442 000	2 099 534	61.0%	81.33%		75%	Replace	1977–82	Yamagata Prefecture	Cd	Japan Energy
		81.12.25 (plan changed)	4 574 140	2 788 570	61.0%	81.33%		75%					
12	Shinbori Deki River	85.3.29	4 387 580	2 674 857	61.0%	81.33%		75%	Replace	1977–86	Miyagi Prefecture	Cd	Tohoku Alps
		78.7.11	530 000	326 480	61.6%	89.2%	69.1%			1979–80			
13	Iwakura	78.9.20	163 000	101 060	62%	100%	62%		Replace	1978–79	Iwakura (Aichi)	Cd	Nagoya Rashi
14	Sasu River	78.12.12	2 406 576	736 949	32.7%	39.8%	82.15%		Replace	1978–84	Nagasaki Prefecture	Cd	Toho Zinc
15	Motosu	79.5.15	1 945 869	1 357 117	69.7%	97%	82%		Replace	1979–90	Gifu Prefecture	Cd	Sumitomo Cement
16	Oyama-Nogi	79.8.1	274 700	82 410	30%	30%	30%		Replace	1979–82	Oyama	Cd	Origin Electric
17	Noshiro	80.1.22	321 247	137 733	42.9%	85.74%		50%	Replace	1979–83	Akita Prefecture	Cd	Dowa Mining
18	Jinzu 1st	80.2.6 (plan changed)	1 783 000	626 368	35.1%	52.7%		66%	Replace	1979–84	Toyama Prefecture	Cd	Mitsui Metal Mining
		84.7.28	2 247 436	885 265	39.39%								
19	Nihazama River	80.4.25	409 000	130 062	31.8%	47.7%		66%	Replace	1980–85	Miyagi Prefecture	Cd	Mitsubishi Metal
20	Watarase River	80.10.1	5 656 983	2 885 062	51%	87.7%		75%	Replace	1980–85	Gunma Prefecture	Cd	Furukawa Mining
21	Seki River	81.9.22	808 044	379 783	47%	75.5%		66%	Replace	1981–87	Kumamoto Prefecture	Cd	Mitsui Metal Mining
22	Kami Inayoshi	82.3.2	215 000	107 500	50%			50%	Replace	1982–84	Chiyoda (Ibaragi)	Cd	Riken Vacuum
23	Kosaka	82.8.31 (plan changed)	617 000	201 870	32.8%	65.5%		50%	Replace	1982–90	Akita Prefecture	Cd, Cu	Dowa Mining
		85.9.27	695 278	213 981	30.8%								

TABLE 8.1. *Continued*

Name	Date	Total Cost (1000 yen)	Entrepreneur's Bearing (1000 yen)	Rate	Ratio of Contribution		Approximate Ratio	Works Period	(Years)	Enforcement	Source	Polluter
24 Shukunobe River	82.12.24	191 000	25 460	13.3%	26.7%		50%	Replace	1982–85	Kawauchi (Aomori)	Cu, Cd	Tanaka Mining
25 Jinzu 2nd	84.1.20	10 940 000	4 309 266	39.39%	59.08%		66%	Replace	1983–94	Toyama Prefecture	Cd	Mitsui Metal Mining
(plan changed)	91.9.4	9 054 865	3 566 711	39.39%	59.08%							
26 Inuyama	84.3.12	993 000	89 216	8.98%	17.9%		66%	Replace	1984–88	Aichi Prefecture	Cd, Cu	Developer
(plan changed)	88.7.11	712 000	63 969	8.98%	17.9%		50%		1984–88			
27 Takahara	85.4.23	607 000	306 333	50.47%	75.7%		66%	Replace	1985–88	Jyuoh (Ibaragi)	Cd	Japan Energy
28 Haseo	86.5.20	691 484	197 117	28.50%	42.8%		66%	Replace	1986–90	Ohita Prefecture	Cd, As	Mitsubishi Metal
29 Odagawa	86.9.24	716 000	212 366	29.66%	73.7%	80.3%	50%	Replace	1986–92	Tottori Prefecture	Cu	Japan Energy
30 Kami-iwatsu	87.10.5	256 710	85 612	33.35%	44.46%		75%	Replace	1987–90	Asako (Hyogo)	Cd	Mitsubishi Metal
31 Nishi-inabe	87.10.30	5 172 100	1 281 000	24.7%	56%	44%		Replace	1987–99	Mie Prefecture	Cd	Onoda Cement
32 Nishikawa Mazawa	90.7.10	140 000	47 082	33.6%	67.26%		50%	Replace	1990–92	Yamagata Prefecture	Cd	Godo Resource
(plan changed)	93.3.30	128 400	43 181	33.6%	67.26%		50%					
33 Yuta Shimomae	91.7.12	668 660	14 576	2.18%	4.36%		50%	Replace	1991–97	Iwate Prefecture	Cd	Tanaka Mining
34 Kurobe	91.11.19	2 936 000	1 957 431	66%	100%		66%	Replace	1991–96	Toyama Prefecture	Cd	Japan Energy
35 Jinzu 3rd	92.2.3	19 291 900	7 599 079	39.39%	59.08%		66%	Replace	1992–2004	Toyama Prefecture	Cd	Mitsui Metal Mining
36 Urakawa	93.11.5	290 000	145 290	50.1%	75.2%		67%	Replace	1994–97	Arao City	Cd	Mitsui Metal Mining
37 Agawa-south	93.11.12	1 214 000	127 349	10.49%	20.9%		50%	Replace	1994–97	Kanzaki (Hyogo)	Cd	Mitsubishi Material
38 Ohmuta	94.10.31	982 000	510 050	51.94%	98%		53%	Replace	1995–98	Fukuoka Prefecture	Cd	Mitsui Metal
39 Kazuno	95.3.17	689 000	244 170	35.4%	71%		50%	Replace	1995–99	Akita Prefecture	Cd	Dowa Mining

Source: Environment Agency, Department of Planning

on to compensate the farmer for damages suffered, and (C) offer indemnity for the reduction of the farmer's rice crop".

All the Japanese Agricultural Land Pollution Prevention Measures were made operative by the Agricultural Land Soil Pollution Prevention Law (see Table 8.1) and, in 1976, the Liberal Democratic Party (LDP) calculated that the restoration expenses for 5000 ha land would exceed ¥50 billion. It was at the beginning of 1974, under financial pressure caused by the recession following the oil crisis of that time and the strong appreciation of the yen after 1973, that the Metal Mining Industries began to take measures to reduce the heavy burden of the increasing expenses they had undertaken for the prevention mining pollution.[15] The first step was taken by a member of the LDP in the Diet on April 4, 1974, who argued for the withdrawal of the Health and Welfare Ministry's claims with regard to the effects of cadmium, for the relaxation of the standard set for contaminated rice, and for the soil restoration measures to be halted.

The LDP's proposal was the start of a series of discussion of the cadmium problem. Many questions were raised in the Diet, and Mr. Takaya Kodama, a journalist, published a paper, entitled "Is Itai-Itai Disease a Case of Fictitious Pollution?" in the February 1975 issue of *Bungei Shunju* (a popular journal), while in April 1976 a report on the cadmium pollution problem prepared by the Policy Research Council of LDP was made public by the Department of the Environment.

Since 1974, the Japanese Mining Association has argued in its annual "Request for the Policy Building of Mining" for: (1) a reconsideration of ". . . the Health and Welfare Ministry's views of the cause of Itai-Itai disease"; and (2) further research into the influence of cadmium on humans; as well as (3) the abolition of the dual regulation of rice contaminated with cadmium, which forbids the circulation of rice contaminated by cadmium at levels of 0.4–1.0 ppm; and (4) a re-examination of the agricultural land soil pollution-prevention measures that relate to heavy metals and a reconsideration of soil restoration. The Japanese Mining Association also argues that it is sufficient to set a safety standard for rice as a human good that would grant permission to blend rice contaminated with cadmium.[16]

When calculating the ratio of costs to be borne by Mitsui Metal Kamioka Mining, the Toyama Local Government calculated the regional characteristics ratio by dividing the agricultural land into two, contaminated land within an alluvial fan and non-contaminated land on the outskirts of the fan, while similarly dividing non agricultural land into two, land contaminated by natural pollution within the alluvial fan and non-contaminated land on the outskirts of the fan. The soil of non-agricultural land within the alluvial fan influenced by the Jinzu River contained a higher percentage of cadmium than did the soil of non-agricultural land on the outskirts of the fan, which can, thus, be interpreted as a regional characteristic. It was later shown, however, that Toyama Prefecture's sample of soil from non-agricultural land contained soil taken from flooded land and contaminated land.[17] According to analysis carried out by Professor Shin Honma of Tokyo Agriculture and Engineering University, there was no significant difference in the concentration of cadmium between the under layer of non-agricultural land on the outskirts of the alluvial fan of the Jinzu River and the upper layer of agricultural land investigated by Toyama Local Government, which thus cancels the implication that concentrations of cadmium in the soil are a regional characteristic.[18]

TABLE 8.2. Compensation, nursing allowance and medical expenses paid by Mitsui Metal Mining

(1000 yen)

Year	Compensation	Nursing Allowance	Medical Expenses
1971	66 000		
1972	2 784 000		
1973	64 000	107 278	70 686
1974	49 000	108 522	106 287
1975	22 000	98 180	103 345
1976	19 000	86 995	127 409
1977	17 000	80 588	144 840
1978	13 000	102 956	144 242
1979	9 000	98 942	130 482
1980	19 000	91 074	135 791
1981	16 000	71 059	117 150
1982	10 000	74 532	136 358
1983	85 000	74 466	146 876
1984	20 000	61 971	142 712
1985	26 000	50 798	122 442
1986	43 000	40 837	128 447
1987	11 000	40 323	130 047
1988	14 000	38 492	126 202
1989	46 000	35 222	114 979
1990	12 000	24 720	48 934
1991	15 000	27 676	56 656
1992	53 000	33 060	49 380
1993	151 000	27 520	68 910
1994	31 000	27 627	78 013
1995	40 000	22 960	88 630
1996	10 000	18 560	80 670
1997	2 000	14 950	
Total	3 647 000	1 459 521	2 599 500

Source: The Council of Itai-Itai Disease

If we suppose that cadmium flows naturally from the mineral zone and accumulates downstream in cultivated land, where the contaminated rice was grown, we can also suppose that rice grown in mining zones throughout Japan will also be contaminated. Because rice was grown within the limit of all the valleys influenced by mining throughout the mining zones, we can suppose that such rice must have been contaminated. According to an investigation carried out by Dr. Kohji Iimura and others, no pollution was found in the untouched natural rivers in the mining area, while a lesser natural content of heavy metals was found in rivers running through many Japanese mining operations.[19]

The local government in Toyama Prefecture calculated that the other small companies ratio, other than for Mitsui Metal Kamioka Mining, would be 20.55%, and the local government excluded this ratio from the entrepreneur's share of the burden of costs. However, the processes of mining, ore dressing, and refining have different

TABLE 8.3. The pollution prevention investment by Kamioka Mining

(1000 yen)

Year	Drainage	Smoke Treatment	Closed Mining	Total
1970	127 358	230 000		357 435
1971	145 909	291 814		437 723
1972	235 807	130 202		366 009
1973	433 972	34 076	25 132	493 180
1974	731 083	218 332	108 989	1 058 404
1975	501 370	122 142	91 430	714 942
1976	163 700	132 040	92 838	388 578
1977	54 558	54 826	70 587	179 971
1978	66 876	29 068	4 935	100 879
1979	273 538	7 026	6 806	287 370
1980	183 248	34 737	47 914	265 899
1981	306 177	24 286	18 740	349 203
1982	247 226	56 466	10 000	313 692
1983	330 270	86 865	3 646	420 781
1984	383 198	72 037	6 136	461 371
1985	321 825	166 439	6 128	494 392
1986	280 201	75 904	2 484	358 589
1987	256 034	338 700	3 602	598 336
1988	286 635	109 866	5 477	401 978
1989	308 807	87 655	5 258	401 720
1990	372 854	176 007	7 939	556 800
1991	593 519	168 091	6 730	768 340
1992	549 175	269 125	5 522	823 822
1993	245 262	50 618	7 863	303 743
1994	273 103	54 703	4 826	332 632
1995	365 712	48 997	7 199	421 908
1996	277 434	101 030	4 219	382 683
Total	8 314 851	3 171 129	554 400	12 040 380

Source: Kamioka Mining 1997

environmental influences on the downstream area, and because the bulk differential flotation process was brought into effect, the effect of the ore-dressing technology on the environment has been decisive.[20] Until 1971, before the adoption of a bulk differential floating process designed to increase the quantity of ore, the ratio for the quantity of crude ore to total period was only 2%. Therefore, the contribution decreed for the non-existent industry was negligible, and could easily be excluded.

In response to such criticism, the local government of Toyama Prefecture slightly increased the burden-bearing ratio of Mitsui Metal Kamioka Mining for soil pollution-prevention measures at the Jinzu River site to 39.39% at the most.

As we can see, the item relating to "the degree of recognition as a cause of pollution" in Article 4, Item 1, of the Law Concerning Companies Bearing of the Cost of the Public Pollution Control Works functioned to reduce the amount of costs borne by the polluter, whether or not the pollution was caused by regional characteristics, natural contaminants, or a non-existent entrepreneur.

4. Measures Against Soil Pollution at the Jinzu River Site

During the summer of 1972, the Itai-Itai disease case was finally settled and the victims made an agreement with Mitsui Metal Mining. Based on the agreement, soil prevention measures came into operation. However, things did not go as smoothly as had been expected. This was because the assessors of the land were obliged to designate an area and make a plan for soil pollution prevention so that they could start soil pollution prevention as a public works measure.

As a result, between 1974 and 1977, more than four attempts were initiated to designate land as soil pollution prevention areas ranging from the first attempt, when 1500 ha was designated, to the third attempt, when over 1000 ha was stipulated for soil dressing. As we have seen, there was a roll-back in the controversy over the cause of Itai-Itai disease and, consequently, the local government of Toyama Prefecture, which had planned the land-use project, drastically changed its plans for paddy fields contaminated by cadmium. The result was a delay in the public works project.

At this point, we need to clarify the roles played in this affair by the local governments, and we must therefore distinguish between the town or city government and the prefectural government. Because these governing bodies have different allegiances, they sometimes act at cross-purposes. The town or city governments, being closer to their constituencies, tend to support the victims both politically and financially. In contrast, prefectural governments are more sensitive to powerful pressure from industry and from the guidelines laid down by the central government; hence, their policies do not always favor the victims.

The first pilot scheme, which was undertaken from fiscal year 1979 to fiscal year 1984, covered a small area of only 90 ha, amounting to approximately ¥2.4 billion business expenses, with the share to be borne by the industry fixed at a rate of 35.15%, approximately equal to ¥0.88 billion. From fiscal year 1983 to fiscal year 1994, further public works projects covering 441 ha, with expenses of ¥12.4 billion, were planned, but due to the revision and the reduction of the project in the middle of its operation, these works were cut down to 356 ha, and to a cost of approximately ¥10 billion, while the ratio of costs to be borne by industry rose slightly to 39.39%, equivalent to ¥4 billion.

A third area of 437 ha was next designated for treatment from fiscal year 1992 to fiscal year 2004, with expenses of ¥25.5 billion, the entrepreneur bearing 39.39% of the burden, equivalent to approximately ¥10 billion, although the converted land, excluding agricultural land, eventually extended to 563 ha.

The reason for the increase in the area of land to be converted in the third selected district was a result of an increase in the area of land to be sold as a site for a housing complex and an industrial plant; this arose because of the problem of a successor of land, whereas land designated for agricultural purposes may no longer be diverted from land already treated with soil pollution-prevention measures (Table 8.4).

Now, although more than 20 years have passed since the Itai-Itai disease case was brought before the court, the local government of Toyama Prefecture and Mitsui Metal Mining have succeeded in postponing any further undertaking of soil pollution-prevention measures, and have also succeeded in reducing both the area designated for soil restoration and the ratio of expenses to be borne by the companies.

The situation with regard to the PPP as it relates to Mitsui Mining is this: on August 10, 1972, Mitsui Metal Mining made an agreement with the victims' organization with

TABLE 8.4. Soil pollution prevention at the Jinzu River

Project	Years	Area (ha)	Cost (billion ¥)	Cost (billion ¥) borne by Mitsui (%)
First	1979–1984	90	2.4	0.9 (35.15)
Second	1983–1994	356	10.0	4 (39.39)
Third	1992–2004	437	25.5	10 (39.39)

TABLE 8.5. Cost borne by Mitsui as compensation for the suffering caused by Itai-Itai disease

Project	Cost[a]
Compensation for disease damage and medical care	¥7.8 billion (¥14.5 billion)
Compensation for spoiled rice fields	¥11.8 billion (¥14.1 billion)
Restoration of polluted rice fields	¥15.0 billion (¥16.0 billion)[b]
Pollution prevention investment	¥12.0 billion (¥18.0 billion)
Total	¥46.0 billion (¥63.0 billion)

[a] Numbers in parentheses refer to the volume in terms of the 1996 yen
[b] These figures include the future cost of a third project for the prevention of soil pollution

regard to the following items: (1) compensation for Itai-Itai disease; (2) the cleanup of soil pollution; and (3) the prevention of further pollution. Compensation paid by Mitsui Metal Mining, based on (1), was ¥3.6 billion overall, equivalent to ¥8.9 billion in 1996, ¥1.4 billion for nursing allowances, equivalent to ¥2.4 billion in 1996, ¥2.4 billion for medical expenses, equivalent to ¥3.2 billion in 1996, and ¥7.5 billion in total, equivalent to ¥14.6 billion in 1996.

Expenses for soil pollution-prevention measures under (2) were, as we have already seen, approximately ¥0.9 billion for the first project, equivalent to approximately ¥1.1 billion in 1996, approximately ¥4 billion for the second project, equivalent to approximately ¥4.6 billion in 1996, and ¥4.9 billion altogether, equivalent to ¥5.7 billion in 1996. Although approximately ¥10 billion has been estimated as the cost of the third project, this is likely to be lowered before the project is completed.

In addition, compensation for the suspension of planting and the reduction of rice production in paddy fields contaminated by cadmium by more than 1 ppm amounted to ¥11.8 billion altogether, equal to ¥14.1 billion in 1996. The total amount of compensation originally promised still exceeds the amount of money that Mitsui Metal Mining has so far paid for soil pollution cleanup measures.

The cumulative total of expenses for the source measures based on the Pollution Prevention Agreement came to ¥8.3 billion for drainage (¥11.8 billion in 1996), ¥3.2 billion for smoke treatment (¥5.0 billion in 1996), and ¥0.5 billion for the treatment of closed and abandoned mines (¥0.8 in 1996), giving a total of ¥12.0 billion (¥17.6 billion in 1996). Additional expenses of an on-the-spot inspection came to approximately ¥0.2 billion (Table 8.2, Table 8.3, Table 8.5).

Many other issues that need to be considered—such as death and damaged health due to Itai-Itai and kidney disease caused by cadmium poisoning—are hardly taken into consideration in these calculations. The question, for instance, of the amount of compensation to pay for cases of Itai-Itai disease mentioned in (1) must take into account the difficult problem of how to estimate an amount that will compensate fundamentally for irreparable damage to health. As for the soil pollution-prevention measures under (2), the principle in the Agreement that ". . . the polluter pays the total amount of expenses necessary for the measures [to be] undertaken" has not been adhered to.

As a result of the reduction in the area designated for restoration from 1500 to 1000 ha and the reduction in the company's burden-bearing ratio to 35% or 39%, Mitsui Metal Mining has been able to get off with an overall burden-bearing ratio of approximately 26%–27%.

Besides, although Mitsui Metal Mining compensated for the suspension of planting and reduced income (¥11.8 billion in total), rice contaminated by cadmium to levels of 0.4–1.0 ppm was purchased by the government.

Source measures under (3) were actually overcalculated because investment related to production was included within what was calculated as pollution-prevention measures expenses, because, under the Special Law for Mining Pollution Control, investment related to the closed and abandoned mines was intended to finance the undertakings. Consequently, the expenses Mitsui Metal Mining incurred for pollution-prevention measures should, as truly burden bearing, be discounted.

If it is considered a corollary that the restoration of land to its original condition entails that the same pollution damage should never be repeated, source prevention measures as well as soil pollution-prevention measures will be two of the main pillars supporting the recovery of land damaged by pollution.

In this regard, Kamioka Mining's source prevention measures, based on the Pollution Prevention Agreement, have achieved a remarkable success, aiming at the natural background level of less than 0.1 ppb cadmium outflow to the downstream area, and Kamioka Mining's efforts should be highly regarded.[21]

As to the actual honoring by Mitsui Metal Mining of the PPP, with regard to (1) health casualties compensation, (2) soil restoration, and (3) source prevention measures, their contribution to the burden-bearing of (2), soil restoration, has been greatly reduced due to the cut-backs in the size of the area to be restored and also in the burden-bearing ratio. This was achieved because, in their application of the Agricultural Land Soil Pollution Prevention Law and Law Concerning Companies Bearing of the Cost of the Public Pollution Control Works, Mitsui Metal Mining was successful in limiting expenses to the minimum when it came to (3), source prevention measures even partially connected to the rationalization of production, and (2), the utterly unprofitable business of soil restoration, and also by managing to cover the resultant shortfall in expenses with public funds.[22]

5. Special Law for Mining Pollution Control in Metal Mining Industries and Metal Mining Corporation

Of the various measures taken against accumulated pollution brought about by the metal mining industry, the most important are a Special Law for Mining Pollution Control in metal mining industries and the Partial Revised Law of the Metal Mining Searching Promotion Corporation, both promulgated in 1973. The background to the laws can be found in the responsibility of closed and abandoned mining sites for the heavy metal pollution caused by cadmium and arsenic (Toroku Mining in Kyushu). Because pollution-prevention measures are designed to obviate environmental disruption caused by metal mining, so it is required that waste water from mine shafts should be dealt with permanently, in line with source measures for filling up mine shafts, covering sedimentary fields with soil, and the planting of trees.

The Special Law for Mining Pollution Control provides that the owner of mining rights has a duty to organize a yearly business program of pollution-prevention measures (which does not entail current use), special facilities, and a reserve fund set aside for pollution prevention, although this does not include a compensation fund. Along with this Special Law, the reserve fund system for pollution prevention in Metal Mining Industries was established to ensure that the reserve is exempt from tax. In response to the Special Laws, the Metal Mining Corporation reorganized the Metal Mineral Searching Promotion Corporation to take the necessary steps to prevent metal mining pollution, with a guarantee of obligation, management, and guidance for the prevention of mining pollution. If it takes advantage of the pollution prevention reserve fund provided for by the Special Law, the Corporation ought to be able to cover the expenses of measures for the closing of shafts, the covering of sedimentary fields with soil, and the cultivation of plants.

The fundamental principles for the business of pollution prevention in used special facilities that had been established under the Special Law were put into effect twice, first for a project from 1973 to 1982, with a budget of ¥46 billion at 1977 prices, and the second in 1983, with a budget of ¥20 billion at 1982 prices; however, neither project was completed, and, consequently, a third project had to be drawn up.

From fiscal year 1975, the Metal Mining Corporation began to make loans to companies bearing pollution prevention expenses, and from the latter half of fiscal year 1978, they also began to make loans toward the running costs of shaft mine drainage treatment measures in the used special facilities.

Mines whose mining rights had expired more than 5 years previously are not subject to the mining pollution prevention order of the Mine Safety Act, and even within 5 years after the expiration of mining rights, due to dissolution or bankruptcy of the company, as many as approximately 6000 closed and abandoned mine workings throughout Japan had already escaped from their entrepreneurial obligation to prevent pollution.[23]

To cope with the closed and abandoned mines with a non-existent executor of obligation for mining pollution prevention, the Pollution Prevention Subsidy System came into effect for the fiscal year of 1971; initially, the government subsidy was two-thirds, but then, from fiscal year 1975, it was increased to three-quarter of the whole subsidy to the Local Government Authority; from fiscal year 1974, shaft water drainage treatment expenses were also eligible for a subsidy. Over fiscal years 1973–1991, the expenses of mining pollution prevention construction amounted to approximately ¥47 billion, approximately ¥35 billion of the ¥47 billion coming from government subsidies, which meant that 126 of 177 mine workings in the process of planning have already completed construction. Over the same period, the closed and abandoned mine workings with an existent executor and an obligation to pay pollution-prevention expenses were financed with approximately ¥17 billion to cover mining pollution prevention construction expenses, the Metal Mining Corporation making a loan of approximately ¥11 billion. (The number of the mine workings due to be constructed was 255.[24])

During the third mining pollution prevention project, designed to run from 1993 to 2002, the mining pollution prevention business is to be financed with ¥4.5 billion annually for mine drainage treatments, which will amount to ¥45 billion over 10 years. In detail, for the approximately 80 mines all told, ¥1.8 billion is to be financed to mines that have a legally obliged person in charge; these include five mines where the pos-towner is independently carrying out mining pollution prevention activities; ¥2.7 billion will go to those mines where there is no legally obliged person in charge.

As we can see, considerable subsidies have been granted to cover closed and abandoned mining pollution measures, in disregard of whether or not the mine is managed by a person who is legally obliged to prevent pollution.

Because metal mine workings have to continue the mine drainage treatment almost permanently, the resultant problems concern who is mainly responsible for this work and who should bear its cost. The Round Table Conference on the Mine Drainage Problem, the private consultative body to the Director of the Site and Environment Bureau of MITI, reported in May 1980 as follows:

> "Given a new interpretation that 'the maintenance and management of the facilities' should be included in the description of 'the necessary facilities' to safeguard them from mining damages, in Article 26, Item 1 of the Mine Safety Act, on the causal basis that 'pollution was caused by the operation of the mining industry', 'expenses to dispose of natural and other sources of pollution' should be borne by another body, that is, the government 'because each mining company's share of burden of mine drainage treatment expenses should be reduced in view of the (national) mineral resources policy', while the local authority's share should also be investigated, since 'the mine drainage treatment is deeply connected to the preservation of the regional environment'."

The decision that other costs for treating "natural and other sources of pollution (other polluter causation)", based on the PPP, ought to be covered by the government was determined "in view of the (national) mineral resource policy", while, at the same time, the local government authority's share of the burden would be determined by adopting the remunerative principle.

In response to the 1980 report, it was decided that, from the fiscal year 1981, the "closed and abandoned mining pollution prevention construction expenses subsidy system" should be applied to natural and other sources of pollution in mine drainage treatment given to the closed and abandoned mine workings managed by a person legally obliged to keep the site clean (the ratio of the central government's subsidy was three-quarters and that of the local government authority one-quarter). In disregard of the calculation formula of the natural and other sources of pollution ratio, a subsidy ratio was, from the beginning, set at one-third of the treatment expenses, which meant that an additional coefficient had to be introduced in order to hold down the expenses within the budget. We may infer from this, half hidden in the background behind this calculation of the natural and other sources of pollution ratio, the nominal ground of the demand for a central government subsidy.

Then, in 1992, the Special Law for Mining Pollution Control was amended for the purpose of establishing fundamental principles for the measures that needed to be taken in mine drainage treatment, an issue that had been raised by the problems of closed and abandoned mine workings. This law established a Mining Pollution Prevention Measures Fund and set up a Designated Pollution Prevention Business Organization, so that the owner of mining rights will, in the first place, be required to donate, for a period of 6 years, money to a fund controlled by the Metal Mining Corporation to cover 50 appropriate closed and abandoned mine workings. With this fund, measures can be taken against pollution. If it is possible to set the rate of interest at 5% in order to secure approximately ¥1.5 million per year to cover mine drainage treatment expenses, a total fund of approximately ¥30 billion will be required. Contributions to the fund are deductible as expenses.

After the completion, in 1998, of these requirements, the designated Mining Pollution Prevention Business Organization, now named the Resource Environmental Center, is to take measures for mine drainage treatment. The Metal Mining Corporation runs the fund. The PPP seems to be behind the idea that the owner of mining rights in closed and abandoned mine workings should make a contribution to such a fund, and that mine drainage measures should be financed by the use of the interest on that contribution, but there are two problems.

First, we are not simply dealing with the cleanup of the environment by the removal of past contaminants, but with indefinitely continuing work for the prevention of mining pollution after the completion of the cleanup operation.

Second, is the question of whether or not the interest on the owner's contribution can correspond adequately to the economical burdens that the measures will impose.

Of approximately 80 Japanese mines that require mine drainage treatment, only approximately 50 can be managed by the fund and the Resource Environment Center, and the remainder of the mines, because there is no company that can be held legally responsible, have nothing to rely on other than subsidies from the government and the local authority (according to the Mining Department of MITI).

As to the issue of subsidies, we should look at many tax incentives established during this period that relate in some way to issues of pollution. The first is the Pollution Prevention Reserve System, in effect from 1972 to 1978, when, in spite of the limited amount of reserve capital, a substantial profit was counted as loss and exempted from tax. In the case of Mitsui Metal Mining, as much as ¥4.4 billion accumulated over the accounting period of March 1975, and was later appropriated for purposes "other than the original ones". In 1967, a Pollution Prevention Equipment Specially Recognized Depreciation System had been established in order to recover invested capital in order to achieve a quick return. And, in 1974, the Metal Mining Pollution Prevention Reserve System, based on the Special Law, was established.

In addition, it was found useful to extend the range of Pollution Prevention Facilities, and so reduce the fixed property tax. From 1971, in case of the re-use of after-treatment contaminated water as water for industrial use, the production facilities were exempted from tax so that almost all production facilities that related to water use were exempted from tax, while from 1978, closed and abandoned mining pollution prevention construction expenses also came to be treated as losses or necessary expenses.

Thus, tax or other financial benefits such as those provided by the Pollution Prevention Reserve, the Metal Mining Pollution Prevention Reserve, the Pollution Prevention Facilities Specially Recognized Depreciation System, and the Reduction of Fixed Property Tax all came to work as indirect or "hidden" subsidies.

6. Conclusion

Japan has taken very strict measures to prevent mining pollution, and some of these measures have, in principle, been so advanced that if they could be effectively carried out there would be a good prospect that the level of pollution in the natural background might be reduced overall.

Therefore, Japanese technology and experience ought to be utilized in areas such as those found in other countries where people live largely on river water and where paddy fields are a characteristic feature of national land use.

However, when we consider the application of the PPP to cases of accumulated heavy metal pollution, we are forced to conclude that because the PPP has been applied only partially, it has been only partially successful in carrying out measures to prevent pollution, to restore polluted soil, and to remove accumulated heavy metal.

For various reasons, such as the distinction that is drawn between the natural and other sources of pollution, the authorities have substituted public money for costs that ought to be paid by the polluter: that is, the central government and local authorities have provided a subsidy.

Yet, as we have seen from the way in which the soil pollution-prevention measures were carried out at Jinzu River site, steps were taken in order to reduce the polluter's share of the costs, and the way in which the natural and other sources of pollution distinction was made did not always match the actual facts of the case.

Because the PPP has not been strictly adhered to, it is unrealistic to try to settle the whole issue of pollution by hoping to apply the PPP alone, without other safeguards and provisions. We badly need a defined principle that will be persuasive enough itself to ensure the removal of accumulated pollution through the payment of expenses out of public money.

Historically, Japan's PPP—which is not quite equivalent to OECD's PPP—has the features of a legal liability. It is therefore easy to combine this principle with the principle of causation and responsibility, and it is embodied in the "degree of recognition as a cause of pollution" clause of the Law Concerning Companies Bearing of the Cost of the Public Pollution Control Works. The Japanese Government has used this principle as a tool to reduce the burden on the polluters by interpreting and applying the principle rather flexibly.

Given this situation, the Government seems to have two agendas: one, to reduce the burden of payment that falls on the polluter, and two, the need to clean up any residual accumulated pollution that is not recognized as the "polluter's responsibility". To cope with the first item, the Government has introduced many schemes to subsidize the costs, including the use of tax. To answer the second need, the Government introduced a resource policy and local government remunerative principle to rationalize the use of public finances.

References (all in Japanese)

1. Miyamoto K (1989) *Kankyo Keizaigaku* [Environmental economics]. Iwanami Shoten, Tokyo, Chapter 4
2. The Water Quality Control Bureau of the Environment Agency and The Agricultural Administration Bureau of the Ministry of Agriculture and Forestry (1972)

Dojyo-Osen-Boshihouno-Kaisetsu [Commentary on the Agricultural Land Soil Pollution Prevention Law]. Chuo Hoki Shuppan, Tokyo, p 22

3. Ibid., p 46
4. Yamanaka S (1970) *Shugiin-Sangyou-Kogai-Taisaku-Tokubetu-Iinkaigiroku* [Parliamentary paper of the 64th Diet of the Lower House, Tasks Force for Industrial Pollution Prevention], Printing Bureau, Ministry of Finance, Tokyo, No 3. p 19
5. The Ministry of Health and Welfare (1970) *Kougaiboushi-Jigyou-niyousuru-Hiyouhutannikansuru-Kenkyu, Chukan-houkoku* [Report of the cost bearing for the public pollution control works, interim report]. Tokyo
6. Yonehara I (1970) *Shugiin-Sangyou-Kogai-Taisaku-Tokubetu-Iinkaigiroku* [Parliamentary Paper of the 64th Diet of the Lower House, Tasks Force for Industrial Pollution Prevention], Printing Bureau, Ministry of Finance, Tokyo, No 3, p 42
7. Tsuruoka H (1970) *Shugiin-Nourin-Suisan-Iinkaigiroku* [Parliamentary paper of the 64th Diet of the Lower House, the Committee of Agriculture, Forestry and Fisheries], Printing Bureau, Ministry of Finance, Tokyo, No 3. p 20
8. *Kougai-Taisaku-Honbu* [The Environmental Pollution Control Headquarters] (1971) *Kougai-Boushi-Jigyohi-Jigyosha-Hutanhouno-Kaisetsu* [Commentary on Law Concerning Companies Bearing of the Cost of the Public Pollution Control Works]. Chuo Hoki Shuppan, Tokyo, p 4
9. The Water Quality Control Bureau of Environment Agency and the Agricultural Administration Bureau of the Ministry of Agriculture and Forestry, op. cit., p 73
10. The Japan Mining Association (1974) *Kozan* [Mining], each volume
11. The Environment Agency (1994) *Kankyo Hakusho* [White paper of environment]. Printing Bureau, Ministry of Finance, Tokyo, p 11
12. (1975) Applied example of the Law Concerning Companies Bearing of the Cost of the Public Pollution Control Works. In: *Kankyo-Kogai-Kankei-Shiryoshu* [Material of environment and public nuisance]. Gyosei, Tokyo, p 647
13. (1976) *Kozan* [Mining] 9:161
14. *Kankyo-Kogai-Kankei-Shiryoshu*, op. cit., p 660
15. Tonegawa H, Yoshida F (1979) *Mitsui Shihon to Itai-Itai Byo* [Mitsui Company and Itai-Itai disease]. Ohtsuki Shoten, Tokyo, p 266
16. (1993) *Kozan* [Mining] 6:5–6
17. Tonegawa H (1980) Problem of cost bearing for soil restoration. *Kogai Kenkyu* [Pollution Studies] 9:39–48
18. Honma S (1982) Problem of regional characteristics of cadmium. *Kogai Kenkyu* [Pollution Studies] 11:60–67
19. Japan Association of Soil and Fertilizer Study (1991) *Dojyo no Yuugai Kinzoku Osen* [Soil pollution by contaminant metal]. Hakuyusha, Tokyo, pp 33–34
20. Tonegawa H, Yoshida F, op. cit., Chapter 4
21. Hata A (1994) *Itai-Itai Byo-Hasseigen-Taisaku-22nenno-Ayumi* [Itai-Itai disease, 22 years of pollution prevention]. Jikkyo Shuppan, Tokyo
22. Global Environment Economy Study Group (1991) *Nippon no Kogai-keiken* [Japanese experience of public nuisance]. Godo Shuppan, Tokyo
23. Yoshida F, Tonegawa H (1978) Historical process of compensation for environmental disruption by mining. *Keizaigaku Kenkyu* [Economic Studies] Hokkaido Univ 28: 73–162
24. Suzuki H (1992) *Shuugiin-Shoko-Iinkai-Kaigiroku* [Parliamentary paper of the 123rd Diet of the Lower House, Committee of Industry and Trade]. Printing Bureau, Ministry of Finance, Tokyo, p 8

Index